Crucians (*Carassius carassius*)

The crucian is a freshwater fish species that has seen a renaissance of interest in angling and conservation communities in Britain. Despite its wide northern European and western Asian distribution, the species is under threat due to multiple causes including the decimation of the small pond habitats upon which it depends. As a small pond specialist, the crucian serves a particularly important role as a charismatic focus encouraging support for the protection and restoration of this highly threatened and much-reduced habitat type and the diversity of other species that it supports.

Crucians (Carassius carassius): *Biology, Ecology, Angling and Culture* takes a comprehensive approach, covering:

- Biology and ecology, including taxonomy
- Angling, including location, baits, methods, and former and current record fish
- Cultural connections, including etymology, artwork and culinary aspects of the fish

These sections are supported by a bibliography and beautiful colour imagery. *Crucians* offers a fascinating read for conservationists, anglers, freshwater biologists and those with interest in the natural world.

W0113219

Crucians
(*Carassius carassius*)
Biology, Ecology, Angling and Culture

Mark Everard

CRC Press
Taylor & Francis Group
Boca Raton London New York

CRC Press is an imprint of the
Taylor & Francis Group, an **informa** business

First edition published 2025
by CRC Press
2385 NW Executive Center Drive, Suite 320, Boca Raton FL 33431

and by CRC Press
4 Park Square, Milton Park, Abingdon, Oxon, OX14 4RN

CRC Press is an imprint of Taylor & Francis Group, LLC

ISBN: 978-1-032-90975-2 (hbk)
ISBN: 978-1-032-90974-5 (pbk)
ISBN: 978-1-003-56079-1 (ebk)

DOI: 10.1201/9781003560791

Typeset in Times LT Std
by KnowledgeWorks Global Ltd.

Contents

About the Author   vii
Introduction   ix

1 What Is a Crucian?   **1**
Physical Features of the Crucian   2
Crucian Habitats and Habits   7
Crucian Superpowers   8
The Natural Geographical Distribution of the Crucian   10
British Crucians   11
The Spread of Crucian   14
The Diet of the Crucian   17
The Crucian Life Cycle   19
Predation and the Shape-Shifting Crucian   22
Crucian Pests and Diseases   24
Crucians and the Taxonomy of the Carp Family   25
More Information About Crucians   37

2 Crucian Fishing   **41**
The Fish and the Fishing   42
Crucian Location   43
The Summer Fish?   44
Crucian Baits   45
Presentation When Crucian Fishing   52
Lure Fishing for Crucians   57
The Fighting Crucian   57
Photographing Crucians   58
Crucian Records in Modern Times   60
Crucian Records in Former Times   62
Crucian Match Catches   64
Crucians as Bait   64
Crucians as Problems   65
The Post-Carp Environment   65
Running a Crucian Fishery   66

3 Crucians and People   **68**
Crucian Etymology   69
Other Meanings of the Word 'Crucian'   69

Crucian Cuisine   72
Crucians in Aquaculture   73
Crucians in Literature   74
Crucians and the Creative Arts   76
Crucians and Nature Conservation   79
Pet Crucians   89
The Economics of Crucians   91
Crucian Societies   92

Bibliography   95
Index   97

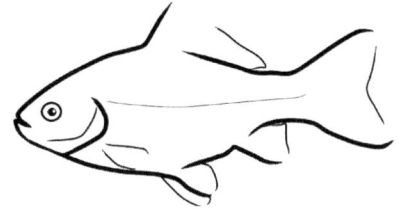

About the Author

Professor Mark Everard is a Visiting Professor at Bournemouth University, as well as Associate Professor of Ecosystem Services at the University of the West of England (UWE Bristol). He also works as a consultant, broadcaster and author. Dr Everard is Vice President of the Institution of Environmental Sciences (IES), a Fellow of the Linnean Society, an Angling Trust Ambassador and a science advisor to WildFish (formerly Salmon & Trout Conservation UK), Tiger Watch (India), Wiltshire Wildlife Trust and a range of other bodies. He is well-known in angling circles as a regular contributor to magazines, radio and television, and as an author of numerous books on fish and fishing. He is also a scientist with a global profile for his research, advocacy and communication on ecosystem management, nature conservation and sustainable development. Dr Everard has published 42 books, 140 peer-reviewed scientific papers as well as frequent magazine, TV and radio contributions. He is often referred to as 'Dr Redfin' in the angling press for his special passion for roach fish. For more about Mark and his work, see www.markeverard.co.uk. Mark is also the designer of the 'Dr Redfin 15 Foot Float Rod'.

Introduction

Back in the early 1960s, the countryside was a far moister and more heterogeneous place. It seemed that almost every field on clay landscapes across the Home Counties and the east of England had a pool or pond of some description. The myriad farm ponds, dew ponds and relic bomb craters extant from the still-recent Second World War were abundant, and were valued by farmers as a source of water for livestock or other uses.

In that now remote time, these diverse water bodies were yet to be substantially expunged from the landscape by the march of industrialised 'agricultural improvements' that saw fields enlarged and hedgerows ripped out to make way for larger mechanised and chemically intensive farming methods, riding roughshod over the network of wetland habitats many of which were ploughed over, encroached or treated as dump sites.

The loss of wildlife – water voles, amphibians, water birds, insects and so much more – has been catastrophic. Its consequences are only just marginally dawning on a world still infatuated by short-term financial returns yet blind to long-term and potentially unrecoverable consequences.

Back in that hazy, simpler time, many of the pools scattered across the rather wetter Home Counties and eastern landscape held a stock of little golden 'coins'. The small, rounded forms of crucians (*Carassius carassius*) were commonly found in this network of small ponds. They abounded in some pools, yet often remained overlooked due to both their cryptic habits and small sizes. But, to the youngster, diligently fishing into dusk in the weedy pools overlooked by most 'serious' anglers, this hidden wealth of 'golden coins' could reveal itself as these charismatic little fishes emerged from hiding to feed.

For many years, crucians were underappreciated, the province of the nostalgic angler who was always pleased when one intercepted their hook bait, though they still attracted the attentions of a few specialists.

Massive changes in land use were to decimate the national wealth of small ponds and other crucian-rich habitats. Worse still, widespread stocking or releases of other fish species has imperilled the very integrity and self-sustaining capacities of the beleaguered crucian. In many ways, crucians have been a 'canary in a coal mine' of the wider ways in which we humans have serially degraded the life-giving qualities of the natural environment.

More recently, there have been strong and sustained waves of renewed interest in crucians in angling circles, with the emergence of conservation strategies and crucian specialist waters. Ironically, this has been followed by evidence suggesting that the crucian may not be an entirely native species within the shores of the British Isles. That said, shifting thinking in nature conservation circles means that this may not in reality matter. Not only does this delicate small water body specialist pose no threat to other characteristic wildlife of these dwindling habitats, but it creates positive landowner and wider societal interests in protection and rehabilitation of threatened ecosystems essential not only for crucians but the wealth of wildlife and societal benefits that they provide.

This book is a celebration of the crucian, bringing together aspects of its ecology, angling interest, and wider cultural associations and points of interest. The scientific aspects are thoroughly researched, but references to scientific publications are presented as footnotes that you can read past or delve into as your interest and inclination lead you. Other references of more general interest are listed in the Bibliography at the end of the book. I have dealt with Latin names of species lightly too, to be read or ignored as you please.

Three chapters follow this introduction respectively addressing:

1. **What is a Crucian?** addresses the biology and ecology of this fascinating fish. Physical features of the fish are described, highlighting those that are particularly important for identification. Close attention is paid to selected features as crucians can be confused with some other closely related species, but also as they face threats from hybridisation with potentially interbreeding species. The life cycle is outlined. Some of the fascinating 'superpowers' of the crucian are reviewed, particularly their variable body shape in response to predation and their tolerance of low-oxygen environments. The wider geographical distribution of the fish is covered, addressing both what is known about its natural range and also where it has been introduced more widely.
2. **Crucian Fishing** addresses approaches to angling for the crucian. A range of relevant methods of angling are covered, starting with location of the fish in different waters. Bait options are covered, along with different approaches to their presentation in different settings and the fishing tackle required to achieve this. Aspects of the seasonality of crucian fishing are touched upon, and there are thoughts on fish care in general and also specific considerations relating to photographing specimen fish. Both modern and former record weight crucians are covered. The chapter concludes with thoughts about the re-emergence of crucian fisheries, and considerations germane to running a crucian fishery.
3. **Crucians and People** covers various aspects of cultural interest. The etymology of the word 'crucian' is considered along with different common names in other languages. The significance of crucians in cuisine as well as aquaculture is reviewed. The occurrence of crucians in literature and the creative arts is covered. Threats to crucians are addressed, before turning to crucians in the context of nature conservation and their role in changing thinking about conservation in a pressurised and changing environment. The potential of crucians as pet fish and aspects of the economics of the fish are considered. The chapter concludes with emerging crucian conservation and recovery initiatives, and the societies that have formed around this charismatic fish.

These chapters are supported by a bibliography, which lists the many scientific and popular sources drawn upon to inform this book and that the reader can follow up for more information.

We are seeing a very welcome new dawn of interest and appreciation of the potential importance of this wealth of 'golden coins' in our landscape, perhaps just in time given the mounting pressures that this charismatic and characteristic little fish is facing.

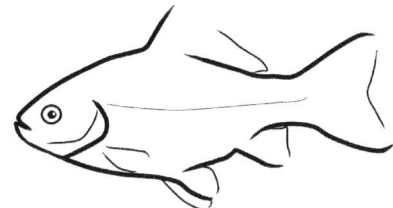

What Is a Crucian?

1

Crucians are members of the carp family, a more detailed description of which is given later in this chapter (Figure 1.1). However, though commonly referred to as 'crucian carp', they are quite different from many of the larger carp species. In his 2023 book *Old Angler Rambling*, Peter Rolfe makes the pertinent comment that

> …*the fish is* Carassius *not* Cyprinus, *related to the Goldfish rather than the Common Carp.*

Peter Rolfe then proceeds to argue that to call this fish a 'carp' is not only misleading (we do not refer to 'goldfish carp') but it can also lead to suffering stigma of the fish through association with the robust and invasive common carp to the extent that

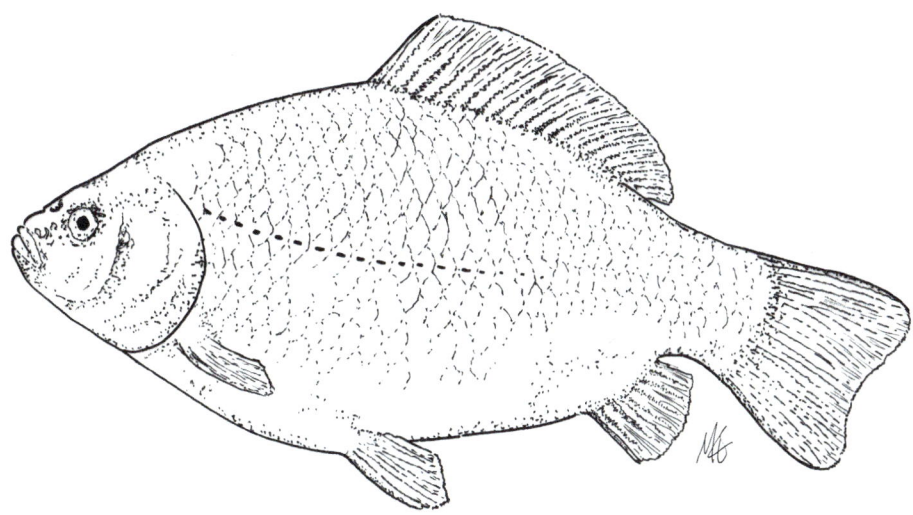

FIGURE 1.1 The crucian: characteristic of small pond habitats. (Image © Mark Everard.)

DOI: 10.1201/9781003560791-1

…stocking with Crucian Carp *is sometimes discouraged because of the false association of the two species.*

Therefore, as a convention, the name 'crucian' is used throughout (except in text quoted from other publications) to avoid confusion with other carps.

Crucians are known by their Latin name *Carassius carassius* (Linnaeus 1758). They were first classified – initially as *Cyprinus carassius* – by the Swedish scientist Carl Linnaeus. Linnaeus, also known as Carl von Linné (1707–1778), is considered the 'father of modern taxonomy' having formalised the Latin binomial nomenclature that is used today. As we shall see, crucians were and still are abundant in the small pools dotting Linnaeus' home nation of Sweden.

Although the name 'crucian', as well as 'crucian carp', is the norm, The Reverend W Houghton wrote in his wonderfully illustrated 1879 book *British Fresh-Water Fishes* that

> *Lacépède calls it the Hamburgh Carp, and some of our Thames fishermen know it by the name of the German Carp. Probably this fish was originally introduced into our own country from Hamburgh, for it is referred to by Linnaeus as being called, in the Transactions of the University of Upsal, by the elder Gronovius* Cyprinus Hamburgher, *as the locality where perhaps it is best known.*
>
> *It is sometimes called the Bream Carp, because the general form of the fish is flat and bream-like.*

Crucians are known by various other common names across their broad geographical range. A wider history of the Latin and common names applied to this fish is covered in Chapter 3 of this book. But the derivation of the name 'crucian' and a couple of other antiquated common names are described by The Reverend W Houghton in *British Fresh-Water Fishes*:

> *The Crucian Carp derives its name from the German word for this fish, namely,* die Karausche, *from whence also the Latin Carassius has been formed.*

PHYSICAL FEATURES OF THE CRUCIAN

Crucians are robust fish, slightly laterally compressed (flattened side to side), though not as extremely as bream species and the flanks are covered in even scales. This fish attains modest size, with the largest recorded length in the scientific literature being 64 centimetres (a little over 25 inches) weighing up to 3.0 kilogrammes (a little over 6.5 pounds). However, crucians principally inhabit small pool environments and their growth is most often limited by circumstances. In most such small-water environments, crucians more typically attain a length of only around 15 centimetres (around 6 inches). Crucians have been recorded as living up to a maximum age of ten years (Figure 1.2).

FIGURE 1.2   A handsome and healthy crucian. (Image © Mark Everard.)

Scientific identification of fish species relies upon a combination of meristics (countable traits such as numbers of scales and of spines and rays in the fins) as well as morphometrics (size and shape using measurable traits such as direct measurements, convex or concave free edges of fins and ratios with other measurements). As we will see later in this chapter, crucians and related fishes have frequently been misidentified. Crucians also exhibit a significant degree of 'phenotypic plasticity' (variable body forms in response to environmental influences). In the light of more recent knowledge, texts that may formerly have been regarded as definitive now have to be treated with some caution. Some key features of the crucian are considered in subsequent text.

Lateral Line and Scales

The crucian's body is covered with even, conspicuous scales. The number of scales along the lateral line – a row of sensory pits through scales generally extending the whole length of the body but sometimes petering out before the tail in the crucian – is a significant meristic feature used in fish identification.

Given the long history of likely misidentification, there is growing consensus that crucians may have a lateral line scale count of 31–34, but that 33 is an ideal count with variance from this requiring further investigation of other features. Various authorities had previously recorded that the number of scales may range from 31 to 36 along a prominent and continuous lateral line running from behind the gill cover to the root of the tail fin along each flank, but this variability often conflated crucians with closely related species and their hybrids. Looking back at many of my photographs of crucians from strains verified by DNA analysis, I concur with the observation of Chris Turnbull in his 2021 book *Willow Pitch VI: Crucian Renaissance*:

> *...the slots along the lateral line of true crucians almost always peter out before the wrist of the tail.*

The number of oblique scales, between the high point of the back and the lateral line but omitting the slotted scales of the lateral line itself, can also be helpful for species identification; there are typically seven oblique scales in case of the crucian.

Colouration

Body colour is variable depending upon habitat and diet, generally shining but ranging from golden-brown to golden-green. A darker reddish-brown or olive-green on the back shades progressively to lighter yellowish or white on the underside. In clearer and lightly weeded waters, the general body colour tends towards golden, but, by contrast, crucians inhabiting densely weeded and dark waters tend to take on a darker body and fin colour (Figure 1.3).

There is also often a rounded dark spot or crescent near the root of the tail fin in young crucians aged up to two years from hatching, though this feature becomes increasingly indistinct in older fish (Figure 1.4).

FIGURE 1.3   Crucians inhabiting densely weeded and dark waters tend to take on a darker body and fin colour. (Image © Mark Everard.)

FIGURE 1.4 Younger crucians, like this 1-year-old juvenile, generally have a distinct spot on the caudal peduncle. (Image © Mark Everard.)

Although body colour has already been noted, colour in general is a highly variable feature and is rarely diagnostic. However, in healthy and particularly clearer water conditions, and especially when the fish are younger or smaller, the pectoral, ventral and anal fins can take on a reddish tinge. In juveniles up to their first year of life, the fins tend to be clear.

Head and Mouth

The head is about one-quarter the length of the body, with a moderately small and delicate terminal mouth. The crucian's mouth is adapted to a diet of fine food particles and is terminal and oriented slightly but not strongly upwards from the lower part of the snout. Neither does the mouth extend into a tube, as is the case with fish such as common bream. Characteristically, the mouth of the crucian lacks barbels (whiskers), a feature significant for distinguishing pure crucians from some of their hybrids (Figure 1.5).

Fin Shape and Numbers of Spines and Rays

The number of spines and rays in the fins, particularly the dorsal and anal fins, is an important meristic feature for distinguishing between fish species and for assessing potential hybridisation, when considered in relation to other features including lateral line count, the shape and orientation of the mouth and the presence or absence of barbels.

FIGURE 1.5 The mouth of the crucian is delicate, oriented slightly upwards and lacks barbels. (Image © Mark Everard.)

Crucians have a long dorsal fin on their backs, held erect at the leading edge by 3–4 spines behind which are 13–22 soft branched rays (technically described as III–IV/13–22). The last (rearmost) dorsal spine is gently serrated. The upper (free) edge of the dorsal fin is convex.

To the rear of the underside of the body, the anal fin of the crucian has 2 or 3 spines and 5–7 (exceptionally 8) soft branched rays (II–III/5–7). Like the dorsal fin, the last anal spine is also gently serrated.

Other characteristic features of the crucian are the lightly forked tail (or caudal) fin, though in many specimens the tail is nearly square with only a slight inward depression. The tail fin lacks spines and is held erect by 18–20 soft rays.

Internal Features

Internally, the spinal column of the crucian comprises 32 vertebrae and there are between 23 and 33 gill rakers (bony or cartilaginous projections from the rear of the gill arches). Members of the carp family have no teeth in the jaw but instead have pharyngeal teeth, located in the throat as the name suggests. These pharyngeal teeth evolved from gill arches and are used to grind food items as they are swallowed. The pharyngeal teeth of the crucian are in a single row.

These internal features require killing and dissecting the fish or, in the case of counting gill rakers, at least anaesthetising it with the risk of damage on examination under the gill covers. They are therefore best left to fish scientists rather than amateurs.

Potential Confusion with Related Species and Hybrids

Later in this opening chapter, we will look at confusion of crucians with related species as well as their hybrids. More details about scale count along the lateral line and the characteristics of fin rays and spines are also considered when distinguishing 'true' crucians from closely related species and hybrids.

CRUCIAN HABITATS AND HABITS

Crucians are fishes of shallow ponds and also of lakes or lake margins that are densely vegetated, offering them cover and refuge from larger competitive or predatory fish species. In his 1983 book *The Complete Freshwater Fishes of the British Isles*, Jonathan Newdick records that

> The Crucian carp is essentially a fish of rich, even over-grown lakes, canals and slow flowing rivers. It tends to spend most of its time near the bottom, can thrive in waters with a low oxygen content, and often becomes torpid in cold weather.

In his 1994 book *Freshwater Fish of the British Isles: A Guide for Anglers and Naturalists*, Nick Giles notes that crucians are amongst an assemblage of fishes typifying a ...

> ... small shallow farm pond used for watering livestock.

Crucians can also prosper in some canals, though generally in abandoned canals where dense vegetation regenerates; significant boat movement tends to result in more open water and less refuge for crucians, which consequently then tend to be displaced by other fish species. Crucians can be found in some slow-moving rivers, though here they are generally restricted to densely vegetated backwaters and oxbows. After observing the crucians in aquaria in my house, I know that even the modest turbulence created when these tanks are topped up with fresh water results in the fish being tumbled; they simply can't handle even modest currents.

I have also found that the crucians in my ponds spend a great deal of the time not just close to the bed but actually swimming around within the layer of soft, unconsolidated sediment coating the bottom. Playing a powerful torch over the ponds at night reveals these fish in open water. But, for the most part of the day, they live not in open water but mobile within the soft, semi-fluid and organically rich upper layer of the pond bed.

This close association with smaller, well-vegetated water bodies is due to the limited capacities of crucians to compete in habitats suiting a range of other fish species, succumbing to the pressures of predation at all life stages and as well as through competition.

CRUCIAN SUPERPOWERS

Nonetheless, crucians possess a remarkable range of 'superpowers' that enable them to survive where many other species cannot. As Bent J Muus and Preben Dahlstrom record in their 1967 *Collins Guide to the Freshwater Fishes of Britain and Europe*,

> *The crucian carp is characteristic of densely overgrown, swampy waters in which often no other fish can exist.*

Of the remarkable robustness of the crucian, The Reverend W Houghton wrote in his 1879 book *British Fresh-Water Fishes* that

> *Like many of the species of this family, the Crucian is eminently retentive of life, and will live for a long time either without water or in water whose impurities would poison other fish; it also manages to exist entirely without food for months, though, as may be supposed in such a condition, it grows very thin.*

The ability of crucians to cope with low-oxygen conditions is enabled by a range of remarkable adaptations. Crucians can abound in small ponds in Finland, Sweden and other Scandinavian countries where oxygen content drops dramatically when ice and snow cover them, preventing exchange of oxygen with the atmosphere and shading aquatic plants preventing photosynthesis. These ponds gradually lose free oxygen throughout extended freezing conditions and vegetation decay, with oxygen concentrations declining close to zero between February and April.

A remarkable experiment entailed exposing crucians to 1–7 days of anoxia (lack of oxygen) followed by 7 days of reoxygenation.[1] Even a short period of anoxia would have been sufficient to kill virtually every other species of freshwater fish, other than those that have the capacity to breathe air (which in the wild would anyway have been prevented by a covering of ice). In the case of the crucian, though, a wide range of biochemical changes occurred. One such change was the triggering of cell growth in the gills, leading to a 7.5-fold increase in respiratory surface area. Other studies[2] have shown that this increase in effective area for oxygen exchange, mopping up any remaining oxygen in the water, is achieved by expansion of the lamellae (thin, plate-like blood-rich outgrowths from the gill arches) substantially increasing the surface area in contact with the water. These morphological changes are triggered by both declining temperature[3] and, reversibly so, depressed oxygen concentrations.[4]

In addition to these changes in gill structure, crucians also undergo a range of biochemical changes as the water temperature declines. One such change is that crucians are prompted to store substantial amounts of glycogen (a polysaccharide of glucose that serves as a form of energy storage) in their brains; the glycogen content of the crucian brain in winter reaches as much as 15 times higher levels than that found in the warmer and more oxygen-rich months of summer.[5] At the same time, the amount of energy required by the brain declines. The combination of these effects enables the crucian's brain to keep functioning throughout winter months when the temperature is low and oxygen may be depleted, extending 150-fold the amount of time, potentially for months, that crucians can survive without oxygen in frigid water. Other organs too have specific adaptations to anoxic conditions and, critically, interactions helping the fish recover when oxygen supply is restored without causing further damage.[6]

In addition to these morphological and physiological adaptations, crucians also substantially reduce their mobility. It can take them a matter of minutes to regain mobility when transferred from cold to warmer water. However, unlike some animals (such as turtles) that become comatose to survive short periods of anoxia, crucians can remain active, albeit at a reduced level, including maintaining brain electrical activity.[7]

A further remarkable physiological adaptation enabling crucians to retain limited brain activity, mobility and other forms of activity in oxygen-free conditions is that glycolysis (conversion of glucose constituting the first step of the respiratory process) can result in the production and excretion of ethanol[8] avoiding a build-up of lactate in the tissues, which is an often-fatal consequence of respiratory processes occurring in the absence of oxygen.[9] By maintaining a degree of activity during anoxia, crucians are also able to seek out oxygen rather than having to wait for it to arrive.

Bent J Muus and Preben Dahlstrom record in their 1967 *Collins Guide to the Freshwater Fishes of Britain and Europe* that

> *The crucian carp can thrive under unfavourable conditions due to its remarkable hardiness. It can endure pollution, oxygen deficiency and winter cold better than most other freshwater fishes. During winter, particularly in the colder parts of its range, it often hibernates almost buried in the mud. In this way it survives the complete freezing of the water, provided the mud around it does not freeze too. Its metabolic processes almost cease and it revives only with the spring thaw.*

Survival by burrowing in mud during extreme winters as well as in the dry season is also recorded by other scientists.[10] As noted previously, my own crucians seem to spend a great deal of the time inhabiting soft, organically rich surface layer of the pool bed in daylight.

Crucians also have a remarkable capacity to survive at both low and high temperatures. They also show a high degree of tolerance of organic pollution. This may possibly be related to their adaptation to oxygen-depleted conditions.

This range of 'superpowers' adapting crucians to withstand the rigours of small pond environments, weathering freeze-ups, deoxygenation and even periodic drying up, enables them to thrive in water bodies that may be inhospitable to other fishes. However, the crucian's adaptations to less hospitable waters come at the cost of enabling them to compete only poorly with other species in more clement waters.

Their adaptations as small pond specialists are, though, problematic given the massive changes in the lowland European landscape that has seen mass elimination of many small ponds and encroachment and/or pollution by agriculture and development on those that remain. Changes in agricultural systems have been a major driver. In the 1960s and sometimes beyond, fields were generally smaller with more hedgerows, each seemingly with a small pool in the corner for the use of farmers and attracting the attentions of children and wildlife alike. These pools were diverse, many serving as water sources for horses and cattle, others dug for marl to spread onto land to increase alkalinity and productivity. But intensifying farming systems demanded bigger fields uninterrupted by hedgerows and ponds. Tractors, unlike the horses they replaced, do not require watering holes, and cheap chemical fertilisers displaced the need and hassle of marl digging and spreading. Today's landscape is more desiccated; many ponds have been infilled by natural processes or deliberately for field enlargement or urban development. Modern landscapes are less resilient to flood and drought and are far less hospitable to wildlife of all types. We are presiding over a truly worrisome biodiversity crisis, especially so for small pond specialists like the once-widespread crucian.

Crucians have more superpowers beside as we will see later in this chapter when considering their phenotypic plasticity (the ability to alter body form in response to environmental changes) in response to predation.

THE NATURAL GEOGRAPHICAL DISTRIBUTION OF THE CRUCIAN

The natural distribution of *C. carassius* is believed to be throughout Europe and Siberia in the western part of Asia.[11] The authoritative website FISHBASE lists (in February 2024) the crucian distribution as:

> *Eurasia: North, Baltic, White, Barents, Black and Caspian Sea basins; Aegean Sea basin only in Maritza drainage; eastward to Kolyma drainage (Siberia); westward to Rhine and eastern drainages of England. Absent from North Sea basin in Sweden and Norway. In Baltic basin north to about 66°N. Widely introduced to Italy, England and France but possibly often confused with* Carassius gibelio.

Günther Sterba stated in his 1959 book *Freshwater fishes of the world* that crucians were

Widely distributed throughout Europe, but absent from Spain, Switzerland, southern Italy and northern Finland.

However, due to their hardiness and unique ability to survive winter anoxia, crucians have been translocated to small lakes and ponds far more widely. As we will see, so have a range of other *Carassius* species, which have been confused with and misidentified as crucians throughout history and certainly before taxonomic distinctions were made. In Northern Europe, as crucians have been used as a food source, we know that they have been widely distributed since medieval times.

Consequently, Bent J Muus and Preben Dahlstrom observe in their 1967 book *Collins Guide to the Freshwater Fishes of Britain and Europe* that

It is difficult to determine the original distribution of the species because for many years it has been introduced and artificially spread to many places, e.g. the British Isles, Spain, France and Norway. It is often confused with the goldfish.

BRITISH CRUCIANS

The truly native status of British fishes, before we humans engaged in their widespread transfer across impassable catchment boundaries, lies in the geological history of the British Isles. Up until the end of the last Ice Age, between 6,500 and 6,200 BCE, mainland Britain's land mass was connected to the European continent by a land bridge known as Doggerland. The Channel River drained this landscape, running southwards to discharge into the Atlantic Ocean. It received the flow of the River Thames from the west and the Rivers Rhine and Scheldt from the east and south in what is now continental Europe. However, a megatsunami was unleashed towards the end of the Ice Age by the breakdown of a gigantic ice lake to the north of Doggerland. This inundated Doggerland, separating what is now the British Isles from continental Europe to the south and east. The name of Doggerland survives now as the Dogger Bank, part of the bed to the south of the North Sea. The lower Channel River now forms the English Channel.

The freshwater fish fauna of England's eastern-flowing rivers, from the Humber system in the north to the Thames in the south, was once common to the continental European rivers formerly mingling their flows in the Channel River. Species with seagoing life stages could spread to the north and west, but those with no such marine phases could not pass dry catchment boundaries. Freshwater fish species such as burbot (*Lota lota*), barbel (*Barbus barbus*) and perch (*Perca fluviatilus*) were thus native only to these western-flowing English rivers. Many fish species have since been widely introduced by humans to waters in other river basins, thriving in many non-native catchments right across the British Isles.

The native or non-native status of crucians in Britain has been a controversial topic. Different authors have expressed opinions, but we also need to look deeper into cases of 'mistaken identity' among crucians and related fish species, particularly among closely related crucian, goldfish and gibel – all three within the genus *Carassius*.

Eric Marshall-Hardy wrote of the crucian in his 1943 book *Coarse Fish*:

> *Asia gave Crucian Carp their original home, whence they were brought to the Continent and thence to these islands.*

As noted previously, The Reverend W Houghton states in his 1879 book *British Fresh-Water Fishes* that

> *Probably this fish was originally introduced into our own country from Hamburgh, for it is referred to by Linnæus as being called, in the Transactions of the University of Upsal, by the elder Gronovius Cyprinus Hamburgher, as the locality where perhaps it was best known.*

However, Houghton expressed differing opinions at the time about distinctions between crucians and gibel:

> *(Carassius vulgaris, var. Gibelio.) THE Prussian Carp, or Gibel Carp, which by Yarrell and some other ichthyologists has been considered a distinct species, is by other authorities, as by Günther, considered merely as a variety of the Crucian.*

Houghton's reference to 'Günther' relates to the 1859-1870 book *Catalogue of the Fishes in the Collection of the British Museum* by Albert Günther. 'Yarrell' is the renowned ichthyologist William Yarrell, who wrote of the Prussian/gibel in his 1859 book *A History of British Fishes in Two Volumes*. Houghton remained uncommitted as to the distinct identity of the gibel, writing that

> *Whether this fish be entitled to rank as a distinct species, I will not pretend to say; at any rate if not specifically distinct, the Prussian Carp is a well marked variety. The Prussian Carp is the Cyprinus gibelio of Bloch, Lacepede, Yarrell, Couch. and other writers; the specific name of gibelio, from the German Giebel, "a gable" or "ridge of a house," would seem to imply that the term was originally given to the Crucian variety, whose back rises abruptly from the head, and forms a prominent ridge; by modern ichthyologists, however, it is now employed to designate the Prussian Carp.*

Perpetuating this lack of distinction between the species, Eric Marshall-Hardy, one-time editor of *Angling* magazine, wrote in his 1943 book *Coarse Fish*

> *THE CRUCIAN OR PRUSSIAN CARP*
> *Chief of these are the absence of barbels and the much less indented caudal or tail fin. The shape and position of the dorsal fin and the general outline of the body are also very different. With regard to the latter, it should be remembered that the body form of Crucian Carp varies considerably, as does that of the common variety. The colour also varies with environment...*

And then we have to tackle potential confusion between crucians and goldfish.

In his 1983 book *The Complete Freshwater Fishes of the British Isles*, Jonathan Newdick records that

> *The Goldfish and the Crucian carp are very similar in appearance; indeed some authorities regard them as varieties of the same species but the Goldfish has a less deep body than the Crucian carp and the serrations on the rays of the dorsal and anal fins of the Crucian carp are far less pronounced than those of the Goldfish. The Crucian carp has 31 to 36 scales along the lateral line ...* (Newdick records that goldfish have 27 to 33 scales along the lateral line.)
>
> *No hybrids involving the Goldfish have been recorded from British waters.*

We also know that the last comment about goldfish not forming hybrids is unfortunately untrue: the free formation of hybrids between naturalised goldfish and crucians is a major threat to the integrity of crucian populations; goldfish also hybridise with other *Carassius* and *Cyprinus* species.

Confusion has persisted even amongst assumed experts: even as late as the 1960s, Bent J Muus and Preben Dahlstrom wrongly stated that the gibel was a wild form of the goldfish.

This detour into historic confusion about crucians, goldfish and gibel is important as it casts doubt upon the natural distribution of the crucian as relayed by many former writers, although it was hardly their fault given the former and still remaining lack of scientific clarity.

In his 1983 book *The Complete Freshwater Fishes of the British Isles*, Jonathan Newdick records that

> *The Crucian carp is locally common throughout England and Wales... was considered to be a fairly recently introduced species until 1975 when the discovery of a Crucian carp bone among Roman remains in London pointed to the fact that it is an indigenous species or at least a very early introduction.*

This discovery of a crucian bone in one Roman midden (a refuse heap generally associated with kitchens) could be indicative of native provenance. However, as only one remain was found, it is more likely that a residue of a crucian or part thereof was brought to Britain by travellers. Dried crucians are still used as food (as we will see in Chapter 3 of this book), and this may have therefore been the case in former times as a means for preservation and transport. Certainly, when reviewing the origins of the British freshwater fish fauna in their 1992 book *Freshwater Fishes*, Peter Maitland and Niall Campbell listed crucians as an introduced species, as does Nick Giles in his 1994 book *Freshwater Fish of the British Isles: A Guide for Anglers and Naturalists*.

Genetic studies on crucians found in England have more recently suggested a medieval (roughly fifteenth century) introduction.[12] This is more or less coincidental with one account of the introduction to Britain of the common carp in the 1817 book *Elements of the Natural History of the Animal Kingdom* by Charles Stewart who stated that the common carp

> *...was introduced into England in 1514, and into Denmark in 1560...*

This is slightly at odds, though perhaps within a margin of error, with the observation by Nick Giles in his 1994 book *Freshwater Fish of the British Isles: A Guide for Anglers and Naturalists* that

> *The Romans were probably responsible for distributing carp widely in European waters. By the reign of England's King Richard II (1377-99) carp were recorded in the royal kitchen.*

Lawrence Wells records in his 1941 book *The Observer's Book of Freshwater Fishes of the British Isles* that the crucian was

> *...introduced to this country from eastern Europe at a later date than the Common Carp and has not been so extensively cultivated...*

Though by no means definitive as to whether this is a primary introduction or that new stock replaced dwindling pre-existing crucian populations, there is now a reasonable degree of confidence in the scientific community that the crucian is not likely to be truly a native fish. However, Peter Rolfe raises some further interesting thoughts in his 2023 book *Old Angler Rambling* that the estimated time of introduction or reintroduction of 500 years based on genetic distancing is necessarily an approximation. Rolfe suggests that it could be closer to 300 years ago, at which time the goldfish was introduced.

We will return in Chapter 3 of this book to the issue of the contested native status of the crucian, and whether or how it actually matters for nature conservation in today's highly modified world.

THE SPREAD OF CRUCIAN

It is certain that humans have been spreading crucians into new waters, into and across Britain and to many other regions, for many centuries. This was aided by the fish's remarkable resilience. Of the ease of transferring crucians, Lawrence Wells wrote in his 1941 book *The Observer's Book of Freshwater Fishes of the British Isles*:

> *...it can live for a long time out of water provided the gills and body are kept moist. This feature, doubtless, enabled them to be transported to this country in days when transport was slow. Wrapped in damp moss they seem to suffer little ill effects even after several hours out of water. Some aquarists, rather than jolt the prize fish about in a can of water and so damage the scales and fins, wrap their pet up in wet newspaper and carry them under their arm.*

The transport of crucians may have initially been for food, and perhaps also for ornamental purposes. More recently, it is also for recreational angling. Crucians may also have spread by means of canals and other human-made connections, or as stowaways when other fish species were stocked into new waters.

Aside from introductions of crucians across mainland Britain, this fish has been widely distributed across the world. We know little of the detail about the deliberate spread of crucians not only throughout England and Wales but also globally. However, one principal route by which many species alien to new lands were introduced was via 'acclimatisation societies' established as voluntary associations in the nineteenth and twentieth century era of colonialism by European settlers. These societies were dedicated to introducing familiar plant and animal species to unfamiliar new environments.

The first recorded acclimatisation society was French, *La Societé Zoologique d'Acclimatation*, founded in Paris in May 1854. Naturally, the British were quick to follow, formally founding the Acclimatisation Society in 1860, not only to actively spread familiar species across their extensive Empire but also naturalising foreign species such as peacocks and pheasant on British soil. The intent of enriching a new region's flora and fauna with species familiar to Europeans was partly one of familiarity, but there was also an occasionally explicit religious undertone of helping the creator fill in the gaps of species He had run out of time to insert into these less-favoured places, benchmarked, of course, against the 'perfect' ecosystems of European homelands! With current knowledge of disastrous consequences – weeds running rife, introduced fishes and birds swamping native species, introduced rabbits overgrazing rangelands in Australia, novel diseases conveyed with introduced species and many more consequences besides – we now know that these practices can be seriously deleterious. But genes, like genies, are harder or impossible to put back into the bottle than to be released.

No Crucians in Ireland

There is no evidence as I write that crucians have been introduced to the island of Ireland. They also appear to be absent from Scotland. As we have seen, crucians were in all probability not a native species but have found their way into England at the hand of humanity, despite the former Doggerland connection.

Their natural absence from Ireland is due to a complete lack of land and river bridges with mainland Britain and Continental Europe. A historical record of the freshwater fishes found in Ireland was created by *Giraldus Cambrensis* ('Gerald of Wales' c.1146–c.1223), a medieval clergyman and chronicler of his times, in his book *The History and Topography of Ireland*. Giraldus Cambrensis recorded that "… *pike, perch, roach, gardon, gudgeon, minnow, loach, bullheads and verones* …" were absent from Ireland. Of the freshwater fishes recorded from Ireland by Giraldus, all are tolerant of salt water either as migratory or brackish water species including brown trout, Atlantic salmon and arctic charr as well as pollan, three-spined sticklebacks, European eels, smelt, shad, three species of lamprey and the (far from common) common sturgeon. These fishes were all able to naturally colonise Ireland's freshwater ecosystems without the aid of humanity due to their salt tolerance.

As we know from the many introductions of coarse freshwater species unable to tolerate marine conditions – common bream, rudd, roach, dace, gudgeon and tench amongst others – Ireland's diverse fresh waters are otherwise ideal. Many introduced species are not only fully established but are also still spreading except into the most inhospitably turbulent environments such as spate rivers.

To date, crucians have not been introduced into Ireland, though would certainly thrive in the diverse weeded still and slow-flowing waters found there. However, caution should always be observed in undertaking species introductions, whether legal or illegal, as they can change the characteristics of natural ecosystems with many knock-on impacts, both ecological and in terms of the many societal benefits that flow from native ecosystems.

India, Australia and Beyond

The British introduced many fish species to India for sport as well as for other reasons. For example, brown trout (*Salmo trutta*) are now well established across the Himalayas and other mountain regions. Tench (*Tinca tinca*) and crucians are known to have been introduced to the Nilgiris (literally 'blue mountains') in the southern Indian state of Tamil Nadu in 1874 as food fish, thus these species spread and became established. Crucians were also bred in a fish farm in the central Indian state of Andhra Pradesh. A study of alien fish distribution in the Himalayas mapped the presence of crucians from 2010 to 2017 across the mountain range, relating this to observed and potential negative impacts arising from introduced non-native fish species and making a case for cessation of stocking with alien species.[13]

Ducks and the Spread of Crucians

An urban myth, often repeated and taken as a truth by many people despite there being not a shred of evidence to support it, is that fish eggs get transferred between waters by being attached to aquatic plants entangled in the feet of ducks. Ducks are streamlined in flight, with weed entanglement unlikely. There has been no study of tangles of weed transported on the feet of ducks, and also no experimental demonstration of the transfer of fish eggs by this route.

Ducks though may, at least hypothetically, have some role in the transfer of fishes between water bodies. The phenomenon of endozoochory, in which resistant biological material can be carried and dispersed within the guts of animals, is well known. Many seeds are dispersed this way, particularly those associated with fleshy and nutritious fruits that are consumed by birds and other animals. Waterbirds are recognised as important for the long-distance dispersal of aquatic plants and aquatic invertebrates with resting eggs or other structures, as well as bryophytes (e.g. mosses, liverworts and hornworts) and also pathogenic and non-pathogenic microbes.[14] Some species of fish too have been found to disperse some types of seeds,[15] and others have been found to play an important role in the passive dispersal of the resting stages of freshwater zooplankton for which migratory species, in particular, may be important dispersal vectors.[16]

Speculation had revolved around the question of whether it was possible for fish eggs to be dispersed in this way. To explore this possibility, captive mallard ducks were fed on vegetation with attached developing eggs of two invasive cyprinid species: common carp and Prussian carp (*Carassius gibelio*).[17] Remarkably, a small proportion (roughly 0.2%) of live embryos could be retrieved from the faeces of these ducks, some subsequently hatching into viable larvae. These studies though are far from conclusive about the actual possibility of fish spreading by endozoochory; the theoretical possibility, including by a *Carassius* species, has never been observed outside of the laboratory. Further study to explore possible bird-mediated colonisation found that 80% of a set of isolated lakes that were newly formed by gravel extraction were colonised by European perch (*Perca fluviatilis*) despite no history of stocking by anglers and managers. This study used multiple lines of evidence – including that perch spawning occurs when waterfowl are in abundance, that the fish release sticky eggs at shallow depths accessible for consumption by or for attachment to waterfowl, and that the appearance of perch tracked waterbird movements – suggesting avian zoochory as a possible dispersal pathway.[18]

Whether dispersal of fish by endozoochory actually occurs in the wild is yet to be proven. The question as to whether it could represent a viable mechanism for the spread of hardy crucians, particularly in the light of their many other adaptations to survival in small pond environments scattered across wider landscapes, remains tantalising but as yet far from definitively answered.

THE DIET OF THE CRUCIAN

Crucians have catholic tastes. They are largely opportunists, making use of a diversity of small plant and animal matter (Figure 1.6). In his 1969 book *The Fishes of the British Isles and North West Europe*, Alwyne Wheeler notes that

> *Little has been recorded concerning the food of this species. It is said to eat both plants and animals, the latter being more important.*

FIGURE 1.6   Bloodworms, the larvae of chironomid midges, live in pond sediments with low oxygen, providing rich and widespread food for many fish species. (Image © Mark Everard.)

Bent J Muus and Preben Dahlstrom describe a rather broader diet of the crucian in their 1967 book *Collins Guide to the Freshwater Fishes of Britain and Europe* stating that

It feeds on plants, insect larvae, especially those of midges and mayflies, and to some extent on planktonic animals.

In his 1983 book *The Complete Freshwater Fishes of the British Isles*, Jonathan Newdick records that

The Crucian carp feeds mainly on invertebrates, especially insect larvae and also on plants.

In his 1943 book *Coarse Fish*, Eric Marshall-Hardy wrote of the crucian that

Both common and Crucian Carp root on the bottom, taking in quantities of mud, together with insects, decaying vegetable matter, shrimps and other crustacea. They feed also on certain water weed, worms and insects, etc., in mid-water, but are, I think, principally vegetarians.

Aspects of this last statement are misleading. The powerful rooting behaviour of common carp, larger fish with far bigger and more powerful protrusible jaws, can be destructive to ecosystems. It is certainly not a behaviour that is fairly attributed to the more delicate feeding habits of the far slighter crucian.

Crucians can certainly make use of very small planktonic food items, which may make presenting angling baits to them challenging during the summer. As Nick Giles wrote in his 1994 book *Freshwater Fish of the British Isles: A Guide for Anglers and Naturalists*, crucians are

...able to feed efficiently on small planktonic crustaceans, for example Daphnia hyalina *when these animals form dense 'blooms' in summer.*

As opportunists, crucians will also feed on available amorphous organic matter. This amorphous matter is known as detritus, muddy-hued and apparently unappealing yet in reality a nutritious food source comprising a wealth of microbes inhabiting or digesting decaying organic matter.

As we have seen, crucians are blessed with remarkable adaptations for surviving in cold and otherwise adverse conditions. These adaptations enable crucians to maintain metabolic activity at a low level, but they become largely inactive in winter. This of course influences their appetite, which diminishes rapidly in cold weather. Nonetheless, various angling authors, including a number writing in Peter Rolfe's excellent 2010 book *Crock of Gold: Seeking the Crucian Carp*, talk of catching crucians in the winter, some even when there is ice on the water. This is not something about which I have personal experience, so I can only repeat what I am told. A summary of quite a lot of reading around this topic suggests that water clarity, often a feature of winter pools devoid of plankton that tends to be restricted to warmer and brighter conditions of spring, summer and autumn, is one of the key features affecting the feeding behaviour. Ever cautious, crucians do not like to venture out into clearer water in bright conditions when predation by fish and birds is most likely. Winter fishing at night targeting crucians close to cover is certainly for the intrepid – to the brave the spoils!

THE CRUCIAN LIFE CYCLE

As denizens of small ponds and the margins of larger still waters, crucians spawn on vegetation. Adult crucians can mature and spawn after their second year (2+ fish). Spawning behaviour typically commences in May or June and can continue, on multiple occasions, to July or sometime later still in warm weather. Bent J Muus and Preben Dahlstrom record in their 1967 book *Collins Guide to the Freshwater Fishes of Britain and Europe* that

> *Spawning continues for some time, as the eggs are shed in at least three portions. Spawning takes place from May-June but is dependent on a minimum temperature of 14°C, the optimum being 19-20°C. The eggs are sticky, light red, 1.5 mm in diameter, and spawned in numbers of 150,000–300,000.*

In his 1983 book *The Complete Freshwater Fishes of the British Isles*, Jonathan Newdick records that

> *Spawning in May and June the Crucian carp lays yellowish or orange eggs measuring 1.5mm in diameter among the vegetation in overgrown areas of shallow water. The eggs adhere to the plants and hatch in six to twelve days when the larvae measure about 5mm. Growth is variable being slow in overgrown or small ponds, but quicker in lakes which have areas of open water. The average length at maturity (between three and four years) is about 10 or 11cm.*

When in spawning condition, female crucians become notably plumper and their bodies remain smooth. Male crucians develop a covering of small, lightly raised and lighter-coloured spawning tubercles around the head and gill covers that are slightly rough to the touch, though they are nowhere near as coarse as those found on male common bream or roach at their spawning time (Figure 1.7). It is believed that these tubercles provide a tactile signal to females during the act of spawning and may help with momentary adhesion when the milt is released. The tubercles disappear after spawning.

At the commencement of spawning, multiple male crucians follow gravid females into dense vegetation, typically in the morning, with no apparent segregation of fish of different sizes. Spawning can be quite boisterous as the males seek to fertilise eggs as they are released by female fish. Eggs released by any individual female may be fertilised by multiple males, a process known as polyandry, maintaining the genetic diversity of the population. There is no evident territorial behaviour; this vigorous activity simply involves competition by attendant males to be near the female rather than expressing aggression to other males.

FIGURE 1.7   Male crucians develop small tubercles around their head and gill covers in advance of spawning. (Image © Mark Everard.)

Studies in different countries have found that female crucians can carry approximately 10,000–250,000 (a quarter of a million) small eggs. These clear or pale-yellow eggs, ranging from 1.4 to 1.7 millimetres in diameter and golden in colour, are sticky, adhering to water plants on release and are fertilised by milt (sperm) released by the attendant males. There is no parental care once spawning has occurred.

Bent J Muus and Preben Dahlstrom comment about crucian eggs and their subsequent development in their 1967 book *Collins Guide to the Freshwater Fishes of Britain and Europe* that

> *They stick to plants and hatch in 5-7 days (c. 100 day-degrees C incubation are required). The young, 4.7-4.9 mm at hatching, have attachment organs in front of the eyes and cling to the plants until the whole of the yolk is consumed in a couple of days.*

Fertilised crucian eggs typically hatch after 4–10 days depending on water temperature. The hatchlings are largely immobile, 4 millimetres (0.15 inches) long and still attached to a yolk sac. After the yolk sac is absorbed after 2–4 days, the larvae become free-swimming having attained a length of 7.0–7.5 millimetres (0.28–0.3 inches).

Painstaking research conducted by my friend and colleague, Dr Adrian Pinder, into the early life stages of coarse fishes was published in 2001 as the still-definitive Freshwater Biological Association *Keys to Larval and Juvenile Stages of Coarse Fishes from Fresh Waters in the British Isles*. Adrian observed and documented fish spawning and the subsequent development in captivity of collected eggs, emerging larvae and growing juveniles, and gives detailed insights into the characteristics of five larval and juvenile development phases. These are summarised as follows:

- Phase 1: There is no vertical band of pigment, but dorsal pigment is in the form of a series of small, punctate melanophores.
- Phase 2: Larvae have a rather pointed, somewhat triangular snout when viewed dorsally, with a snout length to interocular distance greater than 0.60. [Author's note: 'Interocular' means the distance between the eyes.]
- Phase 3: The lateral line lacks pigment and two patches of pigment are on the caudal peduncle, the upper part of the opercula [Author's note: 'opercula' are the gill covers] are covered with numerous tiny melanophores, and the snout is more elongated than common carp at the same stage of development.
- Phase 4: The snout, when viewed dorsally, is more elongated and pointed than the common carp, and there is almost continuous pigmentation in the form of punctate melanophores.
- Phase 5: By this stage, the dorsal fin becomes discernibly convex and no barbels are evident (distinguishing it from the common carp).

Subsequent juvenile growth rate is highly variable, depending on habitat conditions. In many small pond environments, juvenile crucians may only have attained a body length of 5–7.5 centimetres (2–3 inches) after several years. By contrast, larger still waters of good water quality and with sufficient food may enable the crucians to attain this length in two years and to grow on significantly (Figure 1.8).

Bent J Muus and Preben Dahlstrom write in their 1967 book *Collins Guide to the Freshwater Fishes of Britain and Europe*, that

> *Growth depends considerably on the availability of food. Maturity is normally reached in 3-4 years at a length of 8-15 cm. The males mature in their third year, the females a year later, the latter growing faster than the males from the second year, and living longer.*

This observation of age at maturity is at odds with prior authors who suggest that this may occur after two years. Faster-growing young crucians from my own breeding population overwintered indoors in warmer conditions and with regular feeding appeared to join in with breeding after just two years, so environmental conditions may affect development to the point of sexual maturity. In his 1943 book *Coarse Fish*, Eric

FIGURE 1.8 Crucians can be highly prolific in ideal waters, giving rise to large numbers of juveniles like these juveniles from the author's breeding pond. (Image © Mark Everard.)

Marshall-Hardy was more optimistic, and, as it happens, was also inaccurate, about the growth rate of the crucian by stating that

> ...*they spawn in May and June and grow rapidly, attaining a weight of 1 lb in the second year, but they do not attain anything like the proportions of common Carp.*

PREDATION AND THE SHAPE-SHIFTING CRUCIAN

As crucians develop from juvenile into adult fish, they can change quite radically in body shape in response to predation, particularly in reaction to piscivorous fishes. Crucians, being hardly the swiftest of species, seem disproportionately vulnerable to the predatory attentions of cormorants as well as pike and perch. Other predators also have a taste for small fishes of many species, including such opportunistic fish species such as chub and eels. Many piscivorous birds, such as kingfishers, herons and egrets,

FIGURE 1.9 In the presence of predators, crucians can change into a high-backed body form. (Image © Mark Everard.)

also opportunistically feed on a variety of small fishes and would not refuse a crucian, and neither would predatory mammals including otters and mink. In fact, one winter I lost the broodstock crucians in my small garden breeding pond, most likely due to a visit by an otter judging by the pots and stones displaced around the pond's margins.

Another of the remarkable superpowers exhibited by crucians manifests in their response to the presence of predatory fishes. Although fishes of different species come in a diversity of shapes associated with their swimming abilities and anti-predator adaptations, crucians are an extreme example in terms of what is known as 'phenotypic plasticity'. In larger bodies of water, such as lakes that may also be occupied by predators, the body of the crucian can become shorter and notably high-backed (Figure 1.9). This adaptation appears to enable them to avoid being swallowed, no longer so readily fitting into the mouths of at least some of their potential predators.

The presence of predators such as pike (*Esox lucius*) and perch (*Perca fluviatilis*) has been found to induce crucians to develop deeper bodies, with greater muscle mass. An experimental study found that this phenotypic change was triggered by chemical cues from predatory fishes that had been eating crucians, though not by predators that had been eating invertebrates.[19] The study suggested that this change in body shape was activated by odours released from intestinal bacteria metabolising crucian tissues within the predators. Another set of experiments in Finland in which crucians were placed in tanks linked by tubes respectively to a pike, perch fed on midge larvae, perch fed on fish and also an unpopulated tank, resulted in the crucians developing measurably higher backs over three or four months only when connected to pike and perch fed on fish, but not to perch fed on midge larvae.[20] This deeper body form confers lower vulnerability to ingestion by gape-limited predators (predators whose potential food is limited by their mouth size). Crucians with a deeper body form have also been found experimentally to enable higher swimming speed, acceleration and turning rate during predator evasion, compared with shallow-bodied crucians.[21] Observations over forty years by Peter Rolfe, recorded in his 2010 book *Crock of Gold: Seeking the Crucian Carp*, confirm that this transformation in shape can be surprisingly rapid:

> *Response to the presence of, say, pike and perch is remarkably fast, weeks or months rather than months or generations.*

Over time, when away from the presence of predators, the body form of the crucian tends to revert to the shallow-bodied form. One of the advantages of this reversion is that the high-backed and more muscle-dense form tends to have lower physiological investment in parasite and disease resistance, the immune response of shallow-bodied fish being stronger.[22] Crucians with thinner body forms also feed more efficiently.

This response to chemical signals from predators, which tend to be absent from small lakes and ponds with oxygen sags during winter, can result in dense populations of stunted and slender crucians where no or fewer predatory or competitive species can thrive. By contrast, in lakes where other fish species are present, crucian numbers and densities tend to be low, and individuals grow to larger sizes.

Bent J Muus and Preben Dahlstrom also write of these alternative body shapes in their 1967 book *Collins Guide to the Freshwater Fishes of Britain and Europe*, though they relate this principally to diet and stunting reporting that

> *In small ponds with little available food it grows slowly, and develops into a stunted, big-headed form* (forma humilis) *which is often very numerous. With better conditions, in larger lakes where food is abundant, the fish become very deep bodied* (forma gibelio). *Between these two extremes all intermediate conditions exist.*

Whilst these different body forms can result from varying degrees of stunting, use of the term 'gibelio' by Muus and Dahlstrom is unfortunate and too readily confused with gibel, a different though closely related species *C. gibelio*. In fact, Muus and Dahlstrom wrongly state that the gibel is a wild form of the closely related goldfish *Carassius auratus*. Care has to be taken interpreting such older sources of information and terminology.

CRUCIAN PESTS AND DISEASES

A fascinating scientific paper published in 2005[23] studied the parasite communities infesting crucian populations in open lakes coexisting with other fish species and in ponds where extreme conditions prevented other fish species from occurring.

Whilst crucians were found to host a more diverse composition of parasites in lakes, some parasites were present only in low numbers even in these habitats. This suggests a high degree of resistance to infection, and this has been verified in laboratory experiments. Crucians in pond habitats hostile to other fish species hosted very few parasites. Crucians then have a high degree of physiological resistance to some parasites, further reinforced by their capacity to thrive in conditions hostile to other parasite-bearing fishes.

Crucians are susceptible to spring viremia of carp (SVC), a viral disease of fish that particularly inflicts common carp.[24] This disease susceptibility necessitates good biosecurity, particularly in aquaculture systems.

CRUCIANS AND THE TAXONOMY OF THE CARP FAMILY

As noted previously, the modern accepted Latin name for the crucian is *C. carassius* (Linnaeus, 1758).

In taxonomic terms, crucians fall within the Actinopterygii class of ray-finned fishes. Within this class, they are classified in the order Cypriniformes: the carp-like fishes.

Within this broad order of carp-like fishes, the long-established family Cyprinidae (minnows or carps) covered approximately 3,160 species (in 376 genera) of varying forms occurring in the fresh waters (only two species occur in fully marine waters) of North America, Africa and Eurasia. The Cyprinidae were formerly split into 11 quite distinctly different sub-families (see the Box below), one of which was the Cyprininae (the true carps).

THE ELEVEN SUB-FAMILIES OF THE FORMER CYPRINIDAE

- Acheilognathinae (bitterling-like cyprinids)
- Barbinae (barbs)
- Cyprininae (the true carps)
- Danioninae (small, minnow-type fish)
- Gobioninae (the gudgeon-like fishes)
- Leptobarbinae
- Leuciscinae (the dace-like fishes)
- Paedocypridinae (a new sub-family including the world's smallest vertebrate)
- Sundadanioninae
- Tincinae (tench-like fishes)
- Xenocypridinae

Over recent years, there has been growing consensus amongst scientists that the very large, long-established family of minnows and carps was ripe for revision as the former Cyprinidae family was too broad, with many of the sub-families it encompassed in reality comprising a grouping of quite distinct families. A 2018 reclassification of the carps and minnows based on morphological and genetic evidence split the large grouping into several distinct families, many elevating former sub-families into full family status.[25] British examples of these newly redefined families include the revised Cyprinidae (true carps including common carp, barbel and crucians), Leuciscidae (minnows of Europe, Asia and North America including roach, common bream and Eurasian minnow), Tincidae (comprising the sole genus and species *T. tinca*, the tench) and Gobionidae (the gudgeons).

This 2018 reclassification of the former carp and minnow family has gained traction amongst biologists. The revised family of true carps, Cyprinidae, still comprises a wide range of sub-families. One of these is the Cyprininae, which comprises (*Aaptosyax, Carassioides, Carassius, Cyprinus, Luciocyprinus, Procypris* and *Sinocyclocheilus*).

Six *Carassius* species were listed in the online FISHBASE website at the time of writing: *C. carassius* (crucian); *C. auratus* (goldfish); *C. gibelio* (gibel); *Carassius cuvieri* (Japanese white crucian carp); *Carassius praecipuus* (from the upper Nam Chat in central Laos, in the Mekong basin); and *Carassius langsdorfii* (from temperate Japan though this fish is possibly a neotype of *C. gibelio*). However, recent taxonomic developments have distinguished the nigorobuna (*Carassius grandoculis*), formerly considered a sub-species or a synonym of the goldfish (*Carassius auratus grandoculis*), as a discrete species native to Japan. Nigorobuna are significant as they have recently been found not just present but locally proliferating in Britain, further compounding pressures on established crucian stocks.

In reality, the genus *Carassius* comprises a complex of species with blurred definitions, some more genetically distinct than others. Furthermore, no one knows exactly where crucians originated as domestication and fish movements had long preceded taxonomic studies. Many prior 'authoritative' texts also most likely misidentified or lumped together different species. Take, for example, the previously cited assertion by The Reverend W Houghton in his 1879 book *British Fresh-Water Fishes* when describing the crucian: "*Lacépède calls it the Hamburgh Carp, and some of our Thames fishermen know it by the name of the German Carp*". We know that the crucian is not generally a fish of rivers and certainly not of ones in which many larger predatory or competitive fishes are present. Add to that the many conflations of gibel with crucians in both scientific and more general literature.

Possibly, *Carassius* is an elastic genus comprising a continuum of forms across a broad geographical range from western Europe through northern Asia and as far as Japan and adjacent Pacific islands, increasingly muddled through free human-mediated mixing and transfers over many prior centuries. Despite recent advanced in genetic analytical capabilities, distinctions between closely related species in the *Carassius* complex can be a little hazy. Their genetic origins and distinctiveness still remain uncertain. Goldfish have a 1,000-year history of domestication, genetic analysis suggesting that the origins of regional genetic strains may have been substantially influenced by selective breeding over this time including interbreeding with crucians that were formerly generally considered the same species and interbred.[26] One genetic study suggests a potential origin of the *Carassius* genus in China's Yangtze river basin, also highlighting that *Carassius* species are also prone to polyploidy (strains in which the cells have more than one pair of chromosomes) indicative of a high degree of genetic fluidity.[27] Even within a limited geographical range across the islands of Japan and adjacent territories, genetic analysis of goldfish (*C. auratus*) found wide regional differences in lineages though with further of mixing of genetic material through fish transfers.[28] The truth of the origin and true distinctiveness of crucians and other *Carassius* species may never be fully revealed, but it is wise to protect distinctive 'species' or genetic strains where they are known to exist.

There are both similarities and distinctions between crucians and other *Carassius* species found in British waters, including: Goldfish (*C. auratus*), gibel (*C. gibelio*) and also the recent distinction of nigorobuna (*C. grandoculis*) that has been found established in southern Britain. Potential free interbreeding between these closely related *Carassius* fishes adds further complexity, diluting the genetic integrity of strains. Distinguishing pure strains of crucian from other carp-like fishes can be a tricky matter, particularly for people less familiar with these fishes. There is certainly plenty of evidence of crucians having formerly been confused with these other species from the carp family.

Without getting too bogged down in uncertainties in the science, or indeed the philosophical question about when a species is truly distinct from a sub-species within what may be a continuum of variable organisms across a broad geographical range, an overview of the principal features of these other British cyprinids is given in the following boxes. Notes on distinctions of related species and hybrids from crucians – or at least what are regarded as 'true' crucians – are included in these summaries.

Common carp (*Cyprinus carpio*), a clearly distinct cyprinid species within the separate genus *Cyprinus*, has been extensively domesticated and bred over many centuries. In his 2020 book *Casting Shadows: Fish and Fishing in Britain*, Tom Fort notes that the first mention of common carp in Britain comes from the household accounts of the 1st Duke of Norfolk in May 1462. Common carp have since been widely stocked in many still waters from duck ponds to reservoirs and are now widespread in British and European waters, as indeed now across much of the world. These fish are therefore rightly named 'common'. However, they were hard to find in British waters even as recently as when I was a child in the 1960s, when they were regarded largely as the province of specialist anglers though some of us lucked into them from time to time.

I am not a fan of the common carp, calling them 'pigs with fins' in more than one of my books due to the problems I have encountered in my work on aquatic systems around the world into which they have been stocked. This is because they are aquatic analogues to pigs, introduced and farmed for their hardiness and efficiency in converting almost any living material in their environments into body mass. Whilst this is an effective means to produce food and provide big specimens for recreational angling, it comes at major cost to the health of their host ecosystems.

FIGURE 1.10   The common carp is a more robust fish with a long, concave dorsal fin and a protrusible mouth with two pairs of barbels. (Image © Mark Everard.)

Common carp have a long dorsal fin that is concave on the outer free edge. They also have large, protrusible mouths surrounded by two pairs of barbels (Figure 1.10). Common carp are also generally larger and more robust fish. There are typically 33–49 scales along the lateral line, though various cultivated strains of common carp have a reduced number of larger scales (known as mirror carp after the large mirror scales) or no scales at all (leather carp). Crucians rarely fare well when common carp are present at anything beyond low density.

On the matter of hybridisation between crucians and common carp, Eric Marshall-Hardy wrote of the crucian in his 1943 book *Coarse Fish* that

> *There is a confusing hybrid between the two in which the barbels are very small.* To this, he added that, *Chief of these are the absence of barbels and the much less indented caudal or tail fin. The shape and position of the dorsal fin and the general outline of the body are also very different. With regard to the latter, it should be remembered that the body form of Crucian Carp varies considerably, as does that of the common variety. The colour also varies with environment....*

Bent J Muus and Preben Dahlstrom record in their 1967 book *Collins Guide to the Freshwater Fishes of Britain and Europe* that

> *Hybrids between carp and crucian carp, which are intermediate in appearance between the parent species, can be produced by artificial means... The carp x crucian-carp hybrid is barren, but hardy and quick to grow.*

Although common carp and crucians tend to hybridise in the wild, artificially bred hybrids between common carp and crucians, known in recreational angling circles as 'F1 carp' or simply 'F1s', have been widely introduced into fisheries across Britain. F1s grow rapidly up to weights of 2–4 lb (0.9–1.8 kilogrammes), or exceptionally up to 10 lb (4.5 kilogrammes), but with the hardiness of crucians

FIGURE 1.11 The mouth of the crucian is small and lacks barbels, whereas that of the common carp is protrusible with two pairs of barbels. (Images © Mark Everard.)

suiting them to densely stocked fisheries where the water quality may be poor. Stocking is principally for sport, rather than for food.

Crucian–common carp F1s may be sterile or exhibit a lower degree of fertility. They strongly resemble common carp, though sharing intermediate features with the crucian. Significantly, F1s lack the two larger pairs of barbels around the mouth of the common carp, but instead possess two very small barbels that are sometimes barely conspicuous (Figure 1.11). The number of scales along the lateral line of a crucian–common carp hybrid is usually 35 or 36, compared with 31–34 in the crucian and 33–40 in the common carp.

The 'take home' message is that any carp-like fish with a trace of whiskers is not a crucian!

However, it also appears that many F1s assumed to be common carp–crucian crosses may instead be crossed with goldfish or gibel, further muddying the genetic soup entering our waters. Although F1s are assumed to be sterile, some may have a low level of fertility, potentially further threatening the genetic integrity of native fish stocks.

Four fishes in the genus *Carassius* that are now found in British waters – crucians, brown goldfish, gibel and nigorobuna – have many common and overlapping features. Consequently, distinguishing 'true' strains of crucians may be far from intuitive. Confident identification of hybrids between these fishes based on visible external features can be even more tricky, though the features of hybrids are generally intermediate between the parent fish.

Goldfish (*C. auratus*) naturally occur in central Asia, China and Japan. Like crucians, they are small, hardy fish that tolerate poor water quality, adapting them to life in small ponds, marshes, thickly vegetated pools and aquaria. Although the cultivated orange form is most familiar, including the diverse body forms produced for the pond and aquarium trade, goldfish tend to revert to a natural bronze body colour when naturalised in the wild, so are known as 'brown goldfish' (Figure 1.12).

Owing largely to their popularity as pet fish, goldfish have been introduced widely across the world, including into many British water bodies. Adverse ecological impacts have been reported after introductions of goldfish in a number of countries.

The superficial resemblance between goldfish and crucians means that the species are often confused, even by experts. In fact, it is only relatively recently that these species have been recognised not only as distinct, but also that we have become aware of the pervasive extent to which goldfish have become established in British waters and around the world.

In his 1969 book *The Fishes of the British Isles and North West Europe*, Alwyne Wheeler wrote

> *The goldfish is distinguished by being slightly less deep in the body than the Crucian carp; the strong spines in the dorsal and anal fins are both deeply serrated; there are usually five rays in the anal fin; the scales are larger, numbering*

FIGURE 1.12 Wild populations of goldfish typically revert to their natural bronze colouration ('brown goldfish') though can still occur in a variety of colours like this pale fish, noting the lack of barbels, flat or concave free edge of the dorsal fin, and fewer scales along the lateral line compared with a crucian. (Image © Mark Everard.)

twenty-eight to thirty-one in the lateral line. The pharyngeal teeth are in a single row of four each side. It attains 12 in (30.5 cm) and a weight of 2 lb (907 g)... D. III-IV/15-9; A. II-III/5-6; lateral line 28-33.

However, Wheeler did not list gibel as a distinct species (nor indeed the Japanese nigorobuna), so this formula to distinguish European species now needs to be treated with caution.

Gibel (*C. gibelio*) have only relatively recently been distinguished as a separate species based on genetic analysis,[29] as is clear from discussion of the goldfish above. (Following Peter Rolfe's consideration about avoiding confusion early in this book, I am dropping 'carp' from the name, as gibel are a species of *Carassius* and not *Cyprinus*.) This is notwithstanding long-standing recognition of the so-called Prussian carp. This is included in the 1879 book *British Fresh-Water Fishes*, in which The Reverend W Houghton wrote that

> *(Carassius vulgaris, var. Gibelio.) THE Prussian Carp, or Gibel Carp, which by Yarrell and some other ichthyologists has been considered a distinct species, is by other authorities, as by Günther, considered merely as a variety of the Crucian. Whether this fish be entitled to rank as a distinct species, I will not pretend to say ; at any rate if not specifically distinct, the Prussian Carp is a well marked variety.*

Previous ichthyologists – including Marcus Bloch (*Oeconomische Naturgeschichte der Fische Deutschlands. Erster Theil*, 1782), Bernard Germain Lacépède (*Illustrations de Histoire naturelle des poissons*, 1798–1803), William Yarrell (*A History of British Fishes in Two Volumes*, 1859) and Jonathan Couch (*A History of the Fishes of the British Islands*, 1877) – had all recognised the Prussian carp as a distinct species then called *Cyprinus gibelio* (Figure 1.13). As regards this species name, The Reverend W Houghton wrote that

> *...the specific name of gibelio, from the German Giebel, "a gable" or "ridge of a house," would seem to imply that the term was originally given to the Crucian variety, whose back rises abruptly from the head, and forms a prominent ridge ; by modern ichthyologists, however, it is now employed to designate the Prussian Carp.*

A 2022 Environment Agency document described gibel as "*... almost indistinguishable from brown goldfish*".[30]

One of the fascinating but also troublesome features of the gibel is the oddity of how it reproduces. Like other cypriniform fishes, gibel spawn in the springtime,

FIGURE 1.13   Gibel, with notably fewer and larger scales than the crucian and with a silvery appearance. (Image © Mark Everard.)

males and females spawning communally in vegetation. However, whilst male gibel can fertilise the eggs released by female fish, gibel exhibit the phenomenon of 'gynogenesis'. What this means is that sperm is necessary to trigger development of the embryo, but that sperm may come from a range of cypriniform species[31]. This can include sperm from other carps but also from less closely related fish species such as roach, common bream and most likely other species, though DNA in the non-gibel sperm does not fuse with the cell nucleus of the gibel egg. The developing embryo therefore contains chromosomes only from the maternal fish. Effectively, clonal populations can rapidly develop, outcompeting other fishes. Whilst some gibel populations may be exclusively female (and therefore haploid, meaning that they have only the female complement of chromosomes), other populations can contain up to 25% males (which are diploid and hence have the full complement of chromosomes).[32] The implication is that it could take the introduction, or else incautious or accidental release, of only one female fish to lead to a mass invasion of this problematic fish. As Peter Rolfe commented pertinently in his 2023 book *Old Angler Rambling*,

> *…and it only takes one escapee to start a plague.*

It is illegal to transport gibel into this country, though their presence was recorded in British waters in 2021 and a number of established populations are now confirmed. In reality, the gibel may have been in Britain for a far longer time introduced though unidentified introductions, aquarium releases or as stowaways when importing other fishes. Worse still, gibel can freely hybridise with crucians.

The presence of gibel can be easily missed in fish trapping experiments compared to crucians potentially leading to underestimation of invasion by gibel, during which time populations may build up hindering conservation actions.[33] The spread of gibel throughout continental Europe is considered a significant threat to the integrity of crucian populations, making British crucian populations significant in terms of conserving this fish across its established range.

Nigorobuna (*C. grandoculis*) has very recently been confirmed as present in Britain at the time of writing, a new entry into the 'rogues' gallery' of potentially hybridising *Carassius* species. Nigorobuna is native to Japan, from the Lake Biwa basin including its tributaries, linked lakes and irrigation canals in Shiga Prefecture near Kyoto. Here, it is important as a culinary fish. The nigorobuna has only recently been determined to be a species in its own right, rather than a localised sub-species of the goldfish (formerly considered *Carassius auratus grandoculis*) (Figure 1.14).

FIGURE 1.14 Nigorobuna (*Carassius grandoculis)* with notably fewer and larger scales and silvery appearance. (Image © Mark Everard.)

When researching for this book, I went to a pond in southern England – location confidential – as gibel had been confirmed there. The fish I found there, along with common carp and rudd, were prolific and easy to catch. Only later did I look closely at the photographs and realise that they were not in fact the claimed gibel, a fish with which I was familiar from continental Europe. Instead, they were a quite different fish that I identified as *C. grandoculis*. This identification was verified by experts at the Environment Agency. So, we have another alien *Carassius* on our hands in Britain, requiring tight biosecurity or, potentially, eradication!

Quite how likely this fish is to spread more widely and hybridise with crucians is currently unknown, but it would be unwise to allow this to happen.

To simplify a complex matter, Table 1.1 documents key features of each fish. Many of the details in the table are derived from a publication by the Angling Trust, National Crucian Conservation Project (NCCP) and Environment Agency[34] addressing crucians, goldfish, common carp and their hybrids. To this, I have added some details from a subsequent Environment Agency[35] document published in 2022 regarding gibel identification once the presence of the fish in Britain had been recognised. There are far fewer studies addressing the features of nigorobuna, one giving details about gill raker counts and morphology (sizes of various physical features) though not about more obvious physical features such as counts of lateral line scales of the number of spines and rays in the fins,[36] though the fish is often described as being largely the same as the goldfish. Nonetheless, my own observations are that the scales are notable larger and fewer, the mouth strongly upward-sloping and the snout notably sharp. A few additional details are drawn from recent scientific studies. Once you get your eye in, the distinctions become clearer, though there is nonetheless considerable scope for confusion.

Hybridisation with Related Fish Species and the Plague of Back-Crossing

Crucian identification can be tricky at the best of times, with high-backed and low-backed fish ostensibly appearing quite different, not to mention adaptation of colour to environment. This is why we rely on key constant physical features such as the number of scales along the lateral line and of rays and spines in the dorsal and anal fins, helping us ensure confident identification. However, a major confounding factor is that crucians are not only highly fecund, but they also tend to readily hybridise with closely related species, particularly goldfish and gibel from the same genus *Carassius* as well as with common carp, as noted previously. All of these species share the same group spawning behaviour and timing. Hybridisation therefore introduces a whole new level of complexity in precise determination of genetically pure strains of crucians, often producing fish with intermediate albeit variable features. This situation is compounded substantially by recent recognition of the presence of gibel and nigorobuna in Britain; in all probability, they and their hybrids have been here for a much longer period of time but have not been recognised and clearly identified.

Adding further to problems associated with these competitor and potentially hybridising species is the fact that hybrids may not be sterile. Back-crossing with and between fertile hybrids obviously introduces considerable complication. Going into more detail about back-crossing is beyond the scope of this book, but a note of caution has already been sounded about the wisdom of stocking 'F1's as it is far from certain

TABLE 1.1 Common external and internal identification features of crucian, goldfish, common carp, some hybrids, gibel and nigorobuna

		CRUCIAN	FERAL GOLDFISH	COMMON CARP	CRUCIAN × GOLDFISH HYBRID	CRUCIAN × COMMON CARP	GIBEL	NIGOROBUNA
							Described as almost indistinguishable from goldfish	Same as goldfish, but have more gill rakers
EXTERNAL FEATURES								
Lateral line	**Scale count**	32–34	27–29	33–49	29–32	34–36	29–33 (large scales)	27–29 (large scales)
	Description	Often interrupted/ fragmented, sometimes fades towards tail	Continuous, often strong (rarely broken)	Continuous, may be fragmented (mirror carp) or absent (leather carp)	Generally continuous, often strong (sometimes fragmented)	Sometimes present, can be interrupted or complete		Continuous, often strong (rarely broken)
Dorsal fin shape		Convex	Straight/slightly concave	Concave anteriorly with long fin base	Straight or convex (can vary)	Often intermediate of the two	Concave or straight	Straight/slightly concave
Colour	**Pelvic fin**	Orange, often with dark tips	Usually pale, occasionally brown/black	Usually dusky with red tinge	Variable: Dependent on parentage and environment	Variable: Dependent on parentage and environment	Tends to be clear	Usually pale, occasionally brown/black
	Dorsal area	Green/brown	Brown	Bronze/brown	Variable: Dependent on parentage and environment	Variable: Dependent on parentage and environment	Bronze/brown	Brown
	Flanks	Golden bronze	Golden brown	Bronze	Variable: Dependent on parentage and environment	Variable: Dependent on parentage and environment	A pronounced silvery appearance with dull brown tones	Golden brown
	Ventral area	Golden yellow/ orange	Silvery gold	Cream/yellow	Variable: Dependent on parentage and environment	Variable: Dependent on parentage and environment	Cream/white	Silvery gold

(Continued)

TABLE 1.1 *(Continued)*

	CRUCIAN	FERAL GOLDFISH	COMMON CARP	CRUCIAN × GOLDFISH HYBRID	CRUCIAN × COMMON CARP	GIBEL	NIGOROBUNA
Body depth	Laterally compressed	Generally rotund	Generally rotund	Intermediate	Intermediate		Generally rotund
Caudal fin shape	Blunt with shallow fork	Deeply forked (lobes sometimes elongate)	Deeply forked	Forked (lobes sometimes elongate)	Forked	Forked	Deeply forked (lobes sometimes elongate)
Anal fin spine	Lightly serrated	Strongly serrated	Strongly serrated	Strong/moderately serrated	Intermediate	Strong followed by 5 ½ soft rays	Strongly serrated
Dorsal fin spine	Lightly serrated	Strongly serrated	Strongly serrated	Variable: Dependent on parentage		Strongly serrated	
Barbels	Absent	Absent	4 in total (2 in corner of mouth, 2 on top lip)	Absent	Present, very reduced in size and number (2 or 4)	Absent	Absent
Head and mouth shape	Slightly upturned	Central	Central or down-turned, jaws protrusible	Intermediate/ variable	Intermediate/ variable	Pointed snout, upturned mouth	Central
Eye						Silver halo around pupil	
INTERNAL FEATURES							
No. of rakers on first gill arch	21–31	35–43	32–44	38–43	26–32		50–72[37]
Gill raker length	Short	Long	–	Intermediate	Intermediate		

that all F1 specimens are sterile. Furthermore, DNA analysis is now suggesting that some are hybrids of common carp and brown goldfish rather than crucians, further muddying the genetic soup. We should proceed with caution to safeguard our precious genetic heritage.

MORE INFORMATION ABOUT CRUCIANS

If you want to know more about crucian biology, as well as that of other British freshwater fishes, I recommend two of my other books, both also listed in the Bibliography:

- *The Complex Lives of British Freshwater Fishes* (CRC/Taylor and Francis, 2020).
- *Britain's Freshwater Fishes* (Princeton University Press/WildGUIDES, 2013).

In addition, Peter Rolfe's excellent 2010 book *Crock of Gold: Seeking the Crucian Carp* as well as Peter's revised 2024 book *Crock of Gold: The Crucian Revealed* are full of details about this fascinating fish. Chris Turnbull's 2021 book *Willow Pitch VI: Crucian Renaissance* is also well worth consulting.

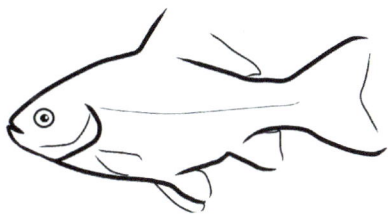

NOTES

1. Sollid, J., Kjernsli, A., De Angelis, P.M., Røhr, A.K. and Nilsson, G.E. (2005). Cell proliferation and gill morphology in anoxic crucian carp. *American Journal of Physiology: Regulatory, Integrative and Comparative Physiology*, 289(4), R1196–1201. DOI: https://doi.org/10.1152/ajpregu.00267.2005.
2. Nilsson, G.E., Dymowska, A. and Stecyk, J.A.W. (2012). New insights into the plasticity of gill structure. *Respiratory Physiology and Neurobiology*, 184(3), pp. 214–22. DOI: https://doi.org/10.1016/j.resp.2012.07.012.
3. Sollid, J., Weber, R.E. and Nilsson, G.E. (2005). Temperature alters the respiratory surface area of crucian carp *Carassius carassius* and goldfish *Carassius auratus*. *Journal of Experimental Biology*, 208, pp. 1109–1116. DOI: https://doi.org/10.1242/jeb.01505.

4. Sollid, J., Angelis, P., Gundersen, K. and Nilsson, G.E. (2003). Hypoxia induces adaptive and reversible gross-morphological changes in crucian carp gills. *Journal of Experimental Biology*, 206, pp. 3667–3673. DOI: https://doi.org/10.1242/jeb.00594.

5. Vornanen, M. and Paajanen, V. (2006). Seasonal changes in glycogen content and Na$^+$-K$^+$-ATPase activity in the brain of crucian carp. *American Journal of Physiology: Regulatory, Integrative and Comparative Physiology*, 291(5), R1482–9. DOI: https://doi.org/10.1152/ajpregu.00172.2006.

6. Dahl, H.-A., Johansen, A., Nilsson, G.E. and Lefevre, S. (2021). The metabolomic response of crucian carp (*Carassius carassius*) to anoxia and reoxygenation differs between tissues and hints at uncharacterized survival strategies. *Metabolites*, 11(7), p. 435. DOI: https://doi.org/10.3390/metabo11070435.

7. Nilsson, G.E. and Renshaw, G.M.C. (2004). Hypoxic survival strategies in two fishes: extreme anoxia tolerance in the North European crucian carp and natural hypoxic preconditioning in a coral-reef shark. *Journal of Experimental Biology*, 207(18), pp. 3131–3139. DOI: https://doi.org/10.1242/jeb.00979.

8. Nilsson, G.E. (1991). The adenosine receptor blocker aminophylline increases anoxic ethanol production in crucian carp. *American Journal of Physiology*, 261, R1057–1060. DOI: https://doi.org/10.1152/ajpregu.1991.261.4.R1057.

9. Nilsson, G.E. and Renshaw, G.M.C. (2004). Hypoxic survival strategies in two fishes: extreme anoxia tolerance in the North European crucian carp and natural hypoxic preconditioning in a coral-reef shark. *Journal of Experimental Biology*, 207(18), pp. 3131–3139. DOI: https://doi.org/10.1242/jeb.00979.

10. Allardi, J. and Keith, P. (1991). *Atlas préliminaire des poissons d'eau douce de France. Coll. Patrimoines Naturels, vol. 4*. Secrétariat Faune Flore, Muséum national d'Histoire naturelle, Paris. 234 pp.

11. Raicu, P., Taisescu, E. and Bănărescu, P..(1981). *Carassius carassius* and *C. auratus*, a pair of diploid and tetraploid representative species (Pisces, Cyprinidae). *Cytologia*, 46, pp. 233–240.

12. Jeffries, D.L., Copp, G.H., Maes, G.E., Lawson Handley, L., Sayer, C.D. and Hänfling, B. (2017). Genetic evidence challenges the native status of a threatened freshwater fish (*Carassius carassius*) in England. *Ecology and Evolution*, 7, pp. 2871–2882. DOI: https://doi.org/10.1002/ece3.2831.

13. Gupta, N. and Everard, M. (2019). Non-native fishes in the Indian Himalaya: an emerging concern for freshwater scientists. *International Journal of River Basin Management*, 17 (2), pp. 271–275. https://doi.org/10.1080/15715124.2017.1411929.

14. Green, A.J., Lovas-Kiss, Á., Reynolds, C., Sebastián-González, E., Silva, G.G., van Leeuwen, C.H.A. and Wilkinson, D.M. (2023). Dispersal of aquatic and terrestrial organisms by waterbirds: A review of current knowledge and future priorities. *Freshwater Biology*, 68(2), pp. 173–190. DOI: https://doi.org/10.1111/fwb.14038.

15. Horn, M.H., Correa, S.B., Parolin, P., Pollux, B.J.A., Anderson, J.T., Lucas, C., Widmann, P., Tjiu, A., Galetti, M. and Goulding, M. (2011). Seed dispersal by fishes in tropical and temperate fresh waters: the growing evidence. *Acta Oecologica*, 37(6), pp. 561–577. DOI: https://doi.org/10.1016/j.actao.2011.06.004.

16. Battauz, Y.S., José de Paggi, S.B and Paggi, J.C. (2015). Endozoochory by an ilyophagous fish in the Paraná River floodplain: a window for zooplankton dispersal. *Hydrobiologia*, 755, pp. 161–171. DOI: https://doi.org/10.1007/s10750-015-2230-4.

17. Lovas-Kiss, Á., Vincze, O., Löki, V. and Lukács, B.A. (2020). Experimental evidence of dispersal of invasive cyprinid eggs inside migratory waterfowl. *Proceedings of the National Academy of Sciences of the United States of America*, 117(27), pp. 15397–15399. DOI: https://doi.org/10.1073/pnas.2004805117.

18. Garcia, F., Paz-Vinas, I., Gaujard, A., Olden, J.D. and Cucherousset, J. (2023). Multiple lines and levels of evidence for avian zoochory promoting fish colonization of artificial lakes. *Biology Letters*, 19(3), PubMed:36946133. DOI: https://doi.org/10.1098/rsbl.2022.0533.

19. Brönmark, C. and Pettersson. L.B. (1994). Chemical cues from piscivores induce a change in morphology in crucian carp. *Oikos*, 70, pp. 396–402.

20. Holopainen, I.J., Aho, J., Vornanen, M. and Huuskonen, H. (1997). Phenotypic plasticity and predator effects on morphology and physiology of crucian carp in nature and in the laboratory. *Journal of Fish Biology*, 50(4), pp.781–798. DOI: https://doi.org/10.1111/j.1095-8649.1997.tb01972.x.

21. Domenici, P., Turesson, H., Brodersen, J. and Brönmark, C. (2007). Predator-induced morphology enhances escape locomotion in crucian carp. *Proceedings of the Royal Society B: Biological Sciences*, 275(1631). DOI: https://doi.org/10.1098/rspb.2007.1088.

22. Vinterstare, J., Hegemann, A., Nilsson, P.A., Hulthén, K. and Brönmark, C. (2019). Defence versus defence: Are crucian carp trading off immune function against predator-induced morphology? *Journal of Animal Ecology*, 88(10), pp. 1510–1521. DOI. https://doi.org/10.1111/1365-2656.13047.

23. Karvonen, A., Bagge, A.M. and Valtonen, E.T. (2005). Parasite assemblages of crucian carp (*Carassius carassius*) – is depauperate composition explained by lack of parasite exchange, extreme environmental conditions or host unsuitability? *Parasitology*, 131(2), pp. 273–278. DOI: https://doi.org/10.1017/S0031182005007572.

24. Godard, M. and Copp, G. (2012). CABI Datasheet: *Carassius carassius* (crucian carp). CABI. [Online.] https://www.cabidigitallibrary.org/doi/10.1079/cabicompendium.90564, accessed 21 February 2024.

25. Tan, M. and Armbruster, J.W. (2018). Phylogenetic classification of extant genera of fishes of the order Cypriniformes (Teleostei: Ostariophysi). *Zootaxa*, 4476(1), pp. 006–039. DOI: https://dois.org/10.11646/zootaxa.4476.1.4.

26. Chen, D., Zhang, Q., Tang, W. and Zhang, J. (2020). The evolutionary origin and domestication history of goldfish (*Carassius auratus*). *Proceedings of the National Academy of Sciences of the United States of America*, 117(47), pp. 29775–29785. DOI: https://doi.org/10.1073/pnas.2005545117.

27. Liu, X.-L., Jiang, F.-F., Wang, Z.-W., Li, W.-Y., Li, Z., Zhang, X.-J., Chen, F., Mao, J.-F., Zhou, L. and Gui, J.-F. (2017). Wider geographic distribution and higher diversity of hexaploids than tetraploids in *Carassius* species complex reveal recurrent polyploidy effects on adaptive evolution. *Science Reports*, 7, p. 5395. DOI: https://doi.org/10.1038/s41598-017-05731-0.

28. Takada, M., Tachihara, K., Kon, T., Yamamoto, G., Iguchi, K., Miya, M. and Nishida, M. (2010). Biogeography and evolution of the *Carassius auratus*-complex in East Asia. *BMC Evolutionary Biology*, 10, p. 7. DOI: https://doi.org/10.1186/1471-2148-10-7.

29. Kalous, L., Bohlen, J., Rylková, K. Petrtýl, M. (2012). Hidden diversity within the Prussian carp and designation of a neotype for *Carassius gibelio* (Teleostei: Cyprinidae). *Ichthyological Exploration of Freshwaters*, 23(1), pp. 11–18.

30. Environment Agency. (2022). *Prussian Carp – a newly discovered non-native fish*. Environment Agency, 13 May 2022. https://anglingtrust.net/wp-content/uploads/2022/05/Gibel-Carp-Hightlight_Final_13_05_22.pdf.

31. Kottelat, M. and Freyhof, J. (2007). *Handbook of European Freshwater Fishes*. Publications Kottelat, Cornol and Freyhof, Berlin. 646 pp.

32. Keith, P. and Allardi, J. (2001). *Atlas des poissons d'eau douce de France*. Muséum national d'Histoire naturelle, Paris. Patrimoines naturels, 47, pp. 1–387.

33. Thomas, K., Brabec, M., Tapkir, S., Gottwald, M., Bartoň, D. and Šmejkal, M. (2023). Sampling bias of invasive gibel carp and threatened crucian carp: implications for conservation. *Global Ecology and Conservation*, 48, e02718. DOI: https://doi.org/10.1016/j.gecco.2023.e02718.

34. Angling Trust, NCCP and Environment Agency. (2021). *Crucian Carp Field Identification Guide*. Angling Trust. [Online.] https://anglingtrust.net/wp-content/uploads/2021/09/Crucian_ID_-_Detailed_Guide.pdf.

35. Environment Agency. (2022). *Prussian Carp – a newly discovered non-native fish.* Environment Agency, 13 May 2022. https://anglingtrust.net/wp-content/uploads/2022/05/Gibel-Carp-Hightlight_Final_13_05_22.pdf.
36. Suzuki, T., Nagano, H., Kobayashi, T. and Ueno, K. (2005). Morphological characteristics of Nigorobuna *Carassius auratus grandoculis* called Io in Lake Nishino. *Fisheries Science,* 71, pp. 679–681. DOI: https://doi-org.ezproxy.uwe.ac.uk/10.1111/j.1444-2906.2005.01015.x.
37. Suzuki, T., Nagano, H., Kobayashi, T. and Ueno, K. (2005). Morphological characteristics of Nigorobuna *Carassius auratus grandoculis* called Io in Lake Nishino. *Fisheries Science,* 71, pp. 679–681. DOI: https://doi-org.ezproxy.uwe.ac.uk/10.1111/j.1444-2906.2005.01015.x.

Crucian Fishing

2

Following years of decline when the crucian almost faded from the angling scene entirely, there has been a welcome renaissance of interest in the crucian and in crucian fishing as more dedicated waters have been created or restored. Younger anglers may only have seen this recent elevation of the crucian into the specimen angling scene, but the older ones amongst us, at least from southern and eastern Britain, tell tales of an abundance of farm ponds dotting the rural landscape, each with its mysteries including often a wealth of golden crucians (Figure 2.1). Echoing the thoughts of this 'old guard'

FIGURE 2.1 The author with a plump and healthy crucian caught pole fishing. (Image © Mark Everard.)

DOI: 10.1201/9781003560791-2

of crucian nostalgics, Chris Turnbull wrote evocatively in the Introduction to the 2021 book *Willow Pitch VI: Crucian Renaissance*:

> There is something traditional and magical about crucians that has its legacy in many older angler's childhood days.

And, in *The Editor's Pitch*, the introductory chapter of the 2017 Issue One of *The Crucian Chronicles*, Chris Turnbull adds:

> ...while it will have gone unnoticed by many, a gentle revolution has been taking place to save our crucians.

For many of us, the purest form of pursuit of the crucian is with a 'dotted down' float, though modern tactics have taken over for the longer-stay angler intent on banking an elusive giant. There are, in fact, three distinct flavours of crucian fishing: casual, specialist (seeking out larger specimens by design) and match fishing. Each has its adherents and favoured techniques, albeit with quite a bit of overlap.

THE FISH AND THE FISHING

My differing classification of angling approaches for crucians relates to the characteristics of the fish. Crucians are, as we know, generally small fish that thrive best in weedy, smaller pools that are less suitable for other predatory and competitive species.

In many waters holding crucian populations, environmental conditions may impose limitations on maximum size, though the crucians found here may be highly fecund. In mixed fisheries where crucians survive, but their capacities to reproduce may be much reduced or even entirely halted, specimens may grow larger in the absence of competition by myriad of their kin. Then there are crucian populations in protected pools with few other competitive or predatory species where individuals may either grow sufficiently large to attract the interest of specimen anglers, or where they are numerous enough to satisfy match anglers.

Akin to my other books addressing angling for the target species, I start off here by understanding better the ecology of the crucian. Then I follow what the great Richard Stuart Walker, the 'father' of modern specimen angling, had to say, principally revolving around the three sequential steps of location, bait and presentation.

In terms of location, you not only have to know that the fish you desire are present, but also to determine where and when they are feeding or can be induced to feed. Then, it is a matter of identifying a bait upon which they will or can be induced to feed. Only when these vital pieces are in place is it worth thinking about the best technique and appropriate tackle to present your selected bait to the fish, to detect bites sensitively, and to have the capability of playing and safely landing a hooked fish.

CRUCIAN LOCATION

Finding a crucian water body can be a matter of following advice from friends or knowledgeable staff in local tackle shops (increasingly rare these days), recommendations in the media (traditional and social), or some good old-fashioned detective work to track down a hidden or undiscovered water or one that people are keeping quiet about. The National Crucian Conservation Project (NCCP) 'crucian catalogue', discussed in Chapter 3 of this book, is also a good place to start.

Bear in mind that crucians are, by inclination, fishes of small and weedy waters; who knows what mysteries are to be found in unexplored 'small blue dots' on local maps? As a youngster, I spent many hours and days walking or on my bike exploring these 'blue dots'. You never quite know what you may discover: a specimen den, a fast-breeder reactor of small fish or perhaps no fish at all. Whatever you may discover, enjoy the journey! (I am not sure that I knew much about the laws of trespass in those sunny days, but asking the consent of the farmer or landowner is a courtesy.) The opposite end of the spectrum is to buy a permit for a known specimen lake, match water or pleasure venue known to hold crucians.

Even on a big water where crucians are present, or are suspected to be, bear in mind the cryptic and retiring nature of the crucian. Look for any form of structure offering cover: a reed or lily bed, an overhanging bough dipping into the water, even a man-made pontoon. Do not discount reedy and weedy fringes right under your feet, which can often shelter crucians fearful of venturing into open water during daylight. It is here that small suitable baits should be offered at or near the bed.

Like many smaller species, crucians tend to be notably crepuscular in habit (most active at dawn and dusk), particularly in clearer conditions. At these times of low light levels, apparently barren waters can suddenly reveal a richness of crucians. John Bailey wrote in his 1992 book *Fisherman's Valley: Seasonal Tips for Coarse Anglers* about a favourite small farm pond:

> *At dusk and dawn the crucians sometimes appear ravenous and the water fizzes with bubbles as they root feverishly on the bed.*

However, close to cover and in murkier water, they may feed throughout the day. Equally, there are waters, particularly those holding larger specimens attractive to the specialist angler, where crucians are emboldened to feed by night. This may particularly be the case for waters that are sparse in habitat features. The characteristics of different waters vary widely, so discovering the uniqueness of each pool may take a bit of trial and error. You may also seek the advice of venue regulars, though bear in mind that received wisdoms in angling may be hand-me-down, outdated or shaped more by the habits of anglers rather than of the fish.

One of the features of crucians that helps inestimably with their location is a rather endearing habit they have of rushing to the surface, flipping over and flicking their tail as they turn before rapidly descending to the bed again. This behaviour generally signals fish in a feeding or breeding mood so always look out for it and not just in the swim that you are fishing as they may be betraying themselves feeding elsewhere. Periodically scanning the surface of the pond or lake, with or without binoculars, can pay dividends.

When approaching a new water, or a familiar water in which crucians tend to be quite mobile, one of the tips that many specialist anglers advocate is to 'follow the wind', exploring the down-wind end of the water body where food tends to accumulate and where warmer upper water layers may build up.

THE SUMMER FISH?

Back in the 1960s and 1970s, when they were more abundant particularly in the small ponds then dotting rural landscapes, crucians were considered 'summer fish'. This was not a species we pursued or expected to encounter other than in the warmer months, from the Glorious 16th of June (bearing in mind that this was before the closed season was lifted on still waters) until the first frosts.

In fact, this 'rule of thumb' was applied to many species, particularly the once far more exclusive waters holding common carp as well as barbel and tench. These larger cypriniform fishes are now routinely fished for throughout the year – perhaps tench less so – as greater specialist angling attention is paid to them, and as their sometimes episodic feeding behaviour in winter is also better understood. Perhaps the changing climate has played a part in this too. This is not to say that they feed freely in all conditions, but that there may be feeding windows particularly with a slight up-tick in water temperature.

So too the crucian. However, there seem to be waters in which crucians do not seem to feed in winter and others where they can be caught all year round. Water depth has been advanced by some as an explanation for this, as shallower waters cool more rapidly and perhaps also leave fishes more exposed particularly to avian predators.

One Scandinavian website[1] documents fishing successfully for crucians in winter, including under ice, using both small jigs and baited hooks with a preference for using some animal-based baits. The recommendation is to avoid deep water and areas where there is deep silt in which the crucians may bury themselves and become dormant in cold weather. Instead, the recommendation given is to locate areas less than 3 metres deep over a sandy or otherwise less silty bed upon which crucians may remain active. Brighter and warmer conditions are also recommended, and this would seem logical if it contributes to a daily peak in water temperature. That said, the fish in the rather low-resolution photographs on that website appear to have deeply forked tails and also have a suspiciously low number of scales along the lateral line, looking more like gibel or brown goldfish than crucians.

Other British writers, particularly on social media sites, note that their crucians remain active in winter, albeit without the degree of activity encountered in warm weather. Feeding with small pellets is recommended, since the fish may not throw up bubbles as they might in summer, though this may be related to the impact of colder water on the rate of methane formation in the sediment than the slower, less energetic activity of the fish.

So, is the feeding activity of the crucian determined by a combination of temperature, bottom sediment type, food availability, light levels and small diurnal rises in temperature? Will the mid-winter crucian become a common phenomenon as we learn more and adapt our approach as with many innovations in angling? There is so much left to learn!

CRUCIAN BAITS

Many authors describe the predominant food of crucians as invertebrates. In reality, crucians are opportunist omnivores along with many cyprinid fishes. Consequently, many baits can be deployed when fishing for crucians. Furthermore, scattering a few loose offerings of hook bait can help to get them hunting for what is offered on the hook. Maggots, casters, worms, sweetcorn and small cubes of luncheon meat are commonly recommended as effective baits for crucians. Bread is a go-to bait for me when fishing for many species, including crucians, though obviously it is less durable and therefore less well suited for long-stay sessions. Many anglers, particularly specialist anglers, also use pellets and small boilies.

Insect Larvae Baits

Insect larvae are an important part of the natural diet of the crucian, and so it is to these invertebrates that we will first turn. Some are popular and readily procured baits that are effective for crucians.

The most used and convenient insect larva baits are maggots in all their diverse types and colours (Figure 2.2). These different types include run-of-the-mill shop maggots, especially those that are bred to larger sizes (gozzers) and other smaller varieties

FIGURE 2.2   Maggots in all their forms – red maggots are a personal favourite – are a popular, convenient and highly acceptable crucian bait. (Image © Mark Everard.)

(pinkies and squats). Typically, the smaller maggots are used as loose feed, with standard or larger maggots on the hook. Whilst maggots are naturally off-white, they can be dyed into a wide range of colours. Though farmed for the angler, maggots are largely natural creatures and are readily identified by crucians as food, aided by their mobility. A friend of mine who fishes for specimen common carp very often adds maggots to his ground bait no matter what is on the hook, as he believes that carp can hear the maggots moving, attracting their attention. However, some anglers prefer dead maggots, generally achieved by a quick freeze and thaw, as they do not burrow into the sediment. Whatever the type, there is no doubt that crucians like maggots. I prefer red maggots but the choice is personal, and many times crucians are feeding in murky or dark conditions so colour may be less of a factor.

Casters too are a productive bait for crucians and many other fishes. These are the chrysalids of the blowfly, produced as maggots metamorphose into their pupal form. Older, darker-coloured casters are less attractive and also tend to float. Casters certainly have their place in crucian fishing. In fact, some specimen anglers swear by them as the best bait.

However, if the water holds an abundance of small roach, rudd, perch or other tiddlers, maggots and casters can be a liability as these fish will get to these baits long before the shy crucians have a chance. Some anglers use a rubber imitation caster or maggot on the hook in these situations to good effect, with the added benefits that many of these imitations are buoyant and float up to feeding level from the bed.

Another common natural food item of the crucian and many other freshwater fishes is the bloodworm. Bloodworms, familiar to anyone who passes their hands through organically rich mud in their garden pond or fishery, have the appearance of small 'worms' around 1-centimetre long and blood-red in appearance due to a high content of the red pigment haemoglobin, which helps them absorb and store sparse oxygen from their generally anaerobic environments. Though worm-like in appearance, bloodworms are in fact not 'worms' at all but are insects: the larvae of chironomids (non-biting midges). They can be profuse in the bed of the kinds of muddy pools and lake edges favoured by crucians. Though favoured by match anglers for targeting smaller fishes, there is, in reality, little need to be so specialised when targeting crucians. A single bloodworm on a hook is something of a 'needle in a haystack' when seeking large crucians from low-density populations and may attract unwanted attentions from competing smaller fishes in mixed fisheries. The use of bloodworms may though have a place on hard-fished crucian-holding match venues.

Many other insect larvae are used in different branches of angling. These include wasp grubs, leather jackets (larvae of daddy long-legs) and the larvae of mayflies, caddis, stoneflies and other aquatic insects. Adult aquatic insects such as water boatmen may also be used on occasion. However, there seems little purpose in deploying any of these wider larval baits when hunting crucians if maggots or casters in their various forms can do the job just as well.

Other Invertebrate Baits

Other natural invertebrates small enough to engulf are readily taken by crucians and other fishes. These include a variety of species of earthworm (Figure 2.3). Lobworms

FIGURE 2.3   Worms of various species, like these brandlings, make excellent crucian baits. (Image © Mark Everard.)

are Britain's largest earthworm species but, though effective baits for many freshwater fish species, may be a little large for crucian fishing. Smaller worm species that I have succeeded with include *Dendrobaena* (red worms) that commonly occur in leaf litter and brandlings (striped redworms) that thrive in compost and other organically rich waste and that exude a yellowish liquid. But one word of warning about worm fishing for crucians: these fish have small mouths and tend to sip in part of the worm giving a bite but not necessarily with the hook in the mouth. For this reason, I prefer to put halves of small worms on a suitably sized hook.

Crucians also eat other small, meaty baits. Natural freshwater shrimps may be a useful bait but are hard to collect and use. However, fragments of shrimp or prawn, available frozen from supermarkets and taken out in small quantities according to need, are certainly enjoyed by crucians.

Mussels, purchased from supermarkets ready-cooked and removed from their shells either frozen or from the chiller cabinet are effective for tench, common bream, common carp and silver bream, and also work well when presented for crucians. Depending on the size of fish present, larger mussels may need to be cut into smaller and more manageable pieces. Mussels are soft baits, so may be relatively quickly picked apart if many smaller fish are present. Pre-baiting with chopped mussels, periodically checking the mussel or cut section on the hook, can account for crucians, the weight of the mussel making 'dead depth' presentation particularly effective.

Artificial meaty baits

Another meaty bait that is effective for crucians is luncheon meat, cut into bite-sized cubes matched to the size of fish you are after. These can be cut fresh from a newly opened tin or alternatively taken as needed from a bag of pre-cut meat that has been frozen ready for use. When finely diced or grated into finer sizes, luncheon meat can be

FIGURE 2.4   Pellets can be attached to the hook with a bait band tied to a hair, banded direct to the shank of the hook, or soft 'expander' pellets can be hooked directly. Smaller pellets soaked in water can form an attractive ground bait when mixed into a slurry. (Image © Mark Everard.)

added directly as an attractive loose feed or into a carrier such as liquidised bread, dried breadcrumb or commercial ground bait to attract the interests of crucians. Luncheon meat cubes can also be flavoured and/or coloured if you believe that this offers you an advantage.

Fishmeal-based pellets are increasingly popular with anglers targeting a wide range of fish species. Crucians are no exception to the attractiveness of the pellet. Standard fishmeal or krill-based pellets are attractive, with pellets of 4–6 millimetres commonly used on the hook. Often, they are attached with a bait band either directly onto the hook shank or tied into a hair off the back of the hook, though soft 'expander' pellets are possibly better as they can avert accidentally hooking the fish outside of the mouth and potentially causing damage (Figure 2.4). Some new-generation pellets are now being made from insect matter, avoiding the environmental damage associated with overharvesting of small sea fish that are important as a food source for coastal bird populations. Finer pellets, generally softened in water into a mush, can form an attractive feed.

Bread Baits

I freely confess to my irrationally great love of bread as a bait for many freshwater fish species! Crucians are no exception to the lure of bread. Bread is, in essence, simply the soft starchy innards of wheat seeds. Fish encounter many types of seed, swollen by immersion, so bread is in many ways a natural bait. It is no surprise then that it is so commonly accepted by fish, regardless of whether they have been exposed to it previously in bait form.

FIGURE 2.5 Small discs of bread are punched from a fresh slice of bread, slightly compressed and impaled on the hook. (Image © Mark Everard.)

My favoured form of bread on the hook is as flake, torn from a fresh white sliced loaf (Figure 2.5). I have no specific brand loyalty in terms of loaves, but generally pick the cheapest on the shelf as the fish seem to like it best! In my view, one of the virtues of bread in flake form is how light it is, particularly when presented on a light hook (this is one reason why I favour spade-end hooks over eyed hooks) as it is substantially neutrally buoyant. For this reason, it wafts up into the path and mouth of a passing fish without requiring them to up-end themselves to reach the sediment surface.

In the clearest waters of mid-summer, when crucians and other fishes are at their most wary, finer discs of punched bread enable very fine presentation. This is particularly when offered under a fine float presented with a whip or a pole.

Bread paste is less frequently used nowadays, but was a staple bait when I cut my angling teeth in the 1960s. Bread paste is made simply by wetting bread and kneading it, often in a clean cloth, until it attains a smooth and cohesive consistency. Bread paste is also amenable to additions of flavourings, including synthetic substances such as food or commercial bait additives, liquidised pellets, or else cheese, meat pastes or other matter. This paste can also be enhanced with colouration. However, the natural aroma of bread is strongly attractive to fish, albeit not necessarily appearing aromatic to the human nose.

Another of the many virtues of bread is the many ways it can be used to create attractive ground bait or loose feed. In addition to what is on the hook, bread is an excellent loose feed. Reflecting again this behaviour, I tend to liquidise bread as finely as possible, or to blend it in water, such that the loose feed reaches the bed of the pool as a mist rather than in particle form.

Another bready bait that I use a great deal when fishing in India, including for mahseer and snow trout as well as other fishes, is atta (a similar paste is known as ragi in southern India). Atta is simply wheat-based chapatti flour mixed with a little water and kneaded into firm paste. Some Indian anglers mix in a little asafoetida and other 'secret' ingredients. But, for my purposes, I have always found that the bready aroma of the atta is entirely adequate, and it is something to which the fish are well-sensitised as a nature-based food source. Small atta balls work nicely for crucians, and the fact that the chapatti flour is stored dry can be useful when no fresh bait is to hand.

Other Plant and Organic Baits

Many cyprinid fishes readily feed on detritus – amorphous organic matter – particularly as vegetation dies back. Though appearing unattractive to us, detritus derives from decaying plant matter and is rich in microbial life and thus is a nutritious food source. Whilst putting a wad of detritus on the hook in a detritus-filled pond is an unlikely formula for success, a variety of other soft, vegetable-based baits are readily accepted.

Sweetcorn is a plant-based bait that can be highly attractive to crucians, as indeed to many other species of non-predatory fish. The attraction includes not only the sweet flavour, but also the bright colour though many people prefer to add artificial coloration. This works for both tinned sweetcorn as well as frozen corn. Tinned corn has the advantage of being packed in a sweet liquid that instantly adds attraction when mixed with ground bait, whilst frozen corn can be much cheaper to use as only sufficient quantities for the session can be extracted from a large pack stored in a freezer. When fishing with corn, particularly when there is a risk of over-feeding crucians or other small fish, my own preference is to put the corn through a liquidiser to produce a sweet, yellow slurry containing a few fragments of less liquified corn, which is then blended with a little bread, poultry meal or commercial ground bait. This then produces a feed mixed to sufficient consistency whether for throwing as small balls, putting in a method feeder, introducing with a bait dropper or swimfeeder, or else as a slurry introduced into a tight location via a pole pot. A few grains of corn are held back for use on the hook or, when using a smaller hook, to be cut in half for that purpose to present a more manageable or less conspicuous mouthful. If presenting corn using a method feeder or other in-line legering method, it can be worth integrating your piece of corn with a piece of buoyant plastic corn so that the bait wafts up and is more readily seen and engulfed by the fish. This same mixing of real and plastic baits as an attractive 'pop-up' can also be deployed when using maggots, casters, small boilies or other baits (Figure 2.6).

A variety of other plant-based baits can be attractive to crucians. Hemp, well-cooked as is the case for all seed and pulse baits to avoid damaging the fish as the grains swell after being consumed, is considered a highly attractive loose feed keeping crucians and other fish rooting around in the sediment. Stewed wheat is a bait that I have used for roach and dace to great effect, as a loose feed and on the hook, though I admit that I have not done so for crucians though there is no reason why it should not work just as well.

Other plant-based baits that have worked well for other species, including sections of pasta and cooked tares, may also work though I have thus far not felt the need to

FIGURE 2.6   A method feeder is attached in-line, filled with groundbait and with a short link tied to the swivel such that a fish picking up the baited hook will self-hook against the weight of the feeder. (Image © Mark Everard.)

experiment as my main crucian baits work well enough. As the old saying goes, "If it ain't broke, don't fix it!" But I often wander around the tinned and dried food aisles when in the supermarket, and in pet supply shops too, suspecting that the next 'wonder bait' is staring me in the face there on the shelves. Indeed, I have discovered some that have really worked for me when fishing for other species!

Boilies and Other Synthetic Baits

Modern baits abound on the shelves of tackle shops. Prime amongst these are boilies (boiled baits) available in a kaleidoscope of sizes, flavours, colours and buoyancies. Also, a wide range of paste baits is available off those shelves. We have already commented on the small size of the crucian's mouth when compared to many of the species for which boilies are aimed, so choose a small boilie if that is your desired approach. You can also cut a boilie section to suit hook and mouth size. The matters of boilie or paste flavour and colour are yours with which to experiment, my own preference being for the more natural. As noted above when considering presentation with a method feeder, the buoyancy of the bait may also be adapted to meet specific needs either by mixing with part of a plastic bait or else, in the case of boilies, a light microwaving to open up gas pores in the bait itself to increase buoyancy.

These ostensibly synthetic baits are really just concoctions of largely natural substances – egg, milk and other proteins, starches and so on – with colourants and flavourings that may or may not be genuinely synthetic. These baits were primarily designed for durability, withstanding the attentions of smaller species for long enough so that desired larger specimens can find them, and so for longer-stay fishing sessions such as night fishing. Some careful adaptation of approach may be required to tune these baits to the specific crucian fishing situation.

Loose Feed

When feeding the swim for crucians, bear in mind that these are more delicate feeders than larger carp or bream, so avoid a tendency to over-feed. This may not only fill the fish up before they can pick up your hook bait, but it may also attract in competitive species. Tactics such as introducing feed in a very targeted way via a swimfeeder, bait dropper or a pole pot is recommended in tight swims, or where you are trying to lure crucians out from close cover. Rather than 'bomb' the selected swim with a large amount of feed early in a session, a 'little and often' approach is generally best to keep aroma in the feeding zone, to avoid offering too many options other than the bait on your hook, and to avoid drawing in competitive species.

PRESENTATION WHEN CRUCIAN FISHING

We have emphasised the small mouth of the crucian when considering baits. Also, unlike a common carp or common bream, the mouth of the crucian does not extend into an elongated tube to probe sediment and suck up large mouthfuls of matter. Neither do crucians tend often to tip themselves us to delve into the bed; rather, they sip in fragments of food that may be as small as individual 'water fleas'. Many underwater videos are now available online of crucians feeding, and it is striking how, unlike common carp, tench and common bream that tend to grub up the sediment for food whilst roach and rudd rush in and out energetically mopping up dislodged offerings, crucians tend to drift through a cloud of feed mouthing food fragments that waft up in suspension thereby averting the need to upend themselves to delve into the sediment. Crucians are gourmands of the fishy world, rather than swillers of pig troughs! These considerations are all highly germane to presentation.

Smaller crucians in particular – the kind we used to catch as kids in those far-off days when farm ponds abounded – were an endless source of fascination and frustration in equal measure as the float bounced, dipped and dithered with no clear sense of a solid bite. As John Bailey wrote in his 1992 book *Fisherman's Valley: Seasonal Tips for Coarse Anglers*,

> *Typically, the crucian will suck in and blow out an item of food several times to soften it before chewing. This behaviour makes bite indication a nightmare and any attempt at a strike nearly impossible.*

I certainly recall the float dancing without evidently indicating a taking fish back in those distant days of childhood, but I do wonder if this was as much due to the crudeness of our approach as to the behaviour of the fish. After all, unlike the tench and perch as well as roach and rudd that our tactics were suited for detecting when they forage on the pool bed for mouthfuls of food, crucians are far more inclined to sip in fine particles and so betray their hunger in a far more subtle manner. In the modern world, there are two ways of tackling this tendency of crucians to mouth baits apparently indecisively, or perhaps just to be feeding with greater delicacy.

The first of these approaches, and the one I most strongly deploy, is using fine pole floats whether presented by running rod and line or with a pole or whip. I will use a pole float as light as conditions permit, set using a silt plummet such that the bait just kisses the bed. (It is important to use a broad-based silt plummet, or a large shot nipped onto the hook, as a standard plummet may sink into a soft pool bed giving an inaccurate reading of depth.) Also, as with all float-based approaches, it is important to regularly check the depth at which the float is set. This is important as the float my slip on the line or subtle changes in water level, perhaps relating to water backing up as wind blows across a pool's surface, may affect the depth at which the bait is presented.

My bottom 'shot' is invariably a swivel, and from this I prefer just a short link of nylon to a small (size 16) hook such that any mouthing of the bait is transferred instantly to the float. A long hook link can mean that the 'bite' is simply not detected at all, or that the fish has already ejected the bait by the time a float movement is detected. I have enjoyed some considerable success with this approach. As crucians tend to be in or very close to cover by day, this necessarily short-range approach is ideal in many crucian waters as the fish may be even as close a foot or so from the bank if this is where you find a drop-off from a reed fringe or other bed of vegetation. Fish three foot out from this cover and bites may simply not be forthcoming. When fishing this way, I strongly prefer a bait that is light and tends to waft up into suspension, averting the need for crucians to tilt down to pick it up. I also occasionally lift and then lower the rig a few inches – no more than that – such that the hook bait wafts seductively. I also strike with a gentle but firm lift at every movement of the float however subtle, generally a short lift of the rod, pole or whip, often with a successful hook-up but enabling the bait to waft back down without disturbing the fish if the movement is simply a line bite. Many is the time I have fished with people who have simply not seen the bite when I have struck into a fish, such is the subtle habit of the crucian! However, if there are many crucians in the swim, this can result in a lot of missed 'bites' as the fish cavort and collide with the end-tackle, the float bouncing in a fizz of fine bubbles at the water's surface.

The 'lift method' is a float fishing approach that I enjoy very much when tench or silver bream fishing (Figure 2.7). The lift method is centred around a very short link between baited hook and a large shot (I prefer to use a swivel) resting on the bed of the pool and critically balancing a float, the float 'lifting' or even toppling over to lie flat on the water's surface as a fish tips itself up after picking up the bait. Although the shot may be relatively heavy, it is critically balanced with the float and so is neutrally buoyant and so will not be felt by the fish as the bait is mouthed. My own preference is actually to under-shot the float when using the lift method and to draw the line back such that the friction on the bed holds the float down to a small dot. For crucians, the lift method rig has to be substantially more delicate than for other species. I have certainly caught numbers of crucians on the lift method, but I also became aware of its limitations given the crucian's different feeding habit compared with larger fish such as bream and tench that probe into the bed. Crucians instead to tend to sip in items wafting into the lower water column. This is best accommodated by a buoyant or light bait, such as a piece of breadflake, that may waft into the eyeline of a passing crucian. A delicately presented

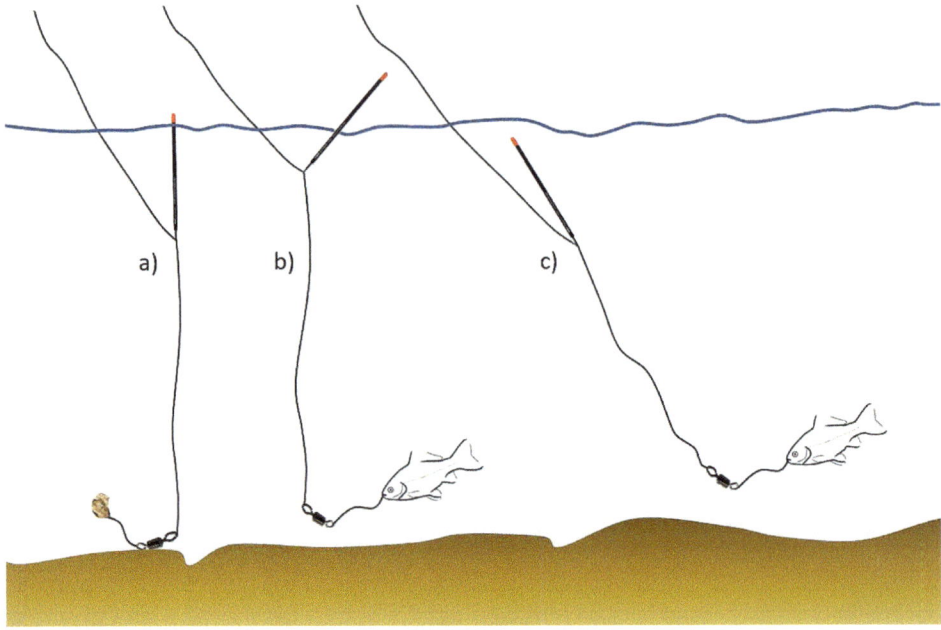

FIGURE 2.7 The lift method: (a) waggler float critically balanced with bottom swivel or shot on bed and short hook link; (b) float lifts if a fish picks up the bait moving the swivel or shot; (c) float sinks as fish moves away. (Image © Mark Everard.)

lift method may be a perfect float-based solution where there is significant wind on the water thwarting more subtle methods, with the line sunk between rod tip and float to avoid surface drag.

A finessed approach that has accounted for many big crucians is to use a fine float set such that the hooked bait is at dead depth. Finely adjusting the depth at which the float is set using a silt plummet or a large shot pinched onto the hook, the aim is to have a bait that is just sufficiently heavy to sink a fine float when it is resting exactly on the bed. Any disturbance at all to the bait as it rests on the bottom inevitably results in movement of the float, be that a lift or disappearance. This is a sure-fire way of seeing and reacting to any movement at all. A further advantage is that the float will rise if a soft bait falls off the hook, be that due to overlong exposure in the water or the attentions of fish. The dead depth method (Figure 2.8) can be used whether pole, whip or waggler fishing, ideally at close range.

Another approach favoured by specimen anglers is to fish 'bolt rig' tactics. The term 'bolt rig' is really a rather crude description of a far more subtle approach, scaled down to match the characteristics of the crucian. This approach is best suited to longer sessions and particularly those running into or through darkness, and is principally the domain of the specialist angler. It entails the use of short hook links of just a couple of inches (about 5 centimetres) extending beyond an in-line swimfeeder or leger weight. Choice of swimfeeder depends on conditions, but a small method feeder is ideal to precisely offer free offerings and a flavour trail around the hook bait. A feeder or lead of at least one ounce is sufficient to ensure that the rig is self-hooking when using fine and

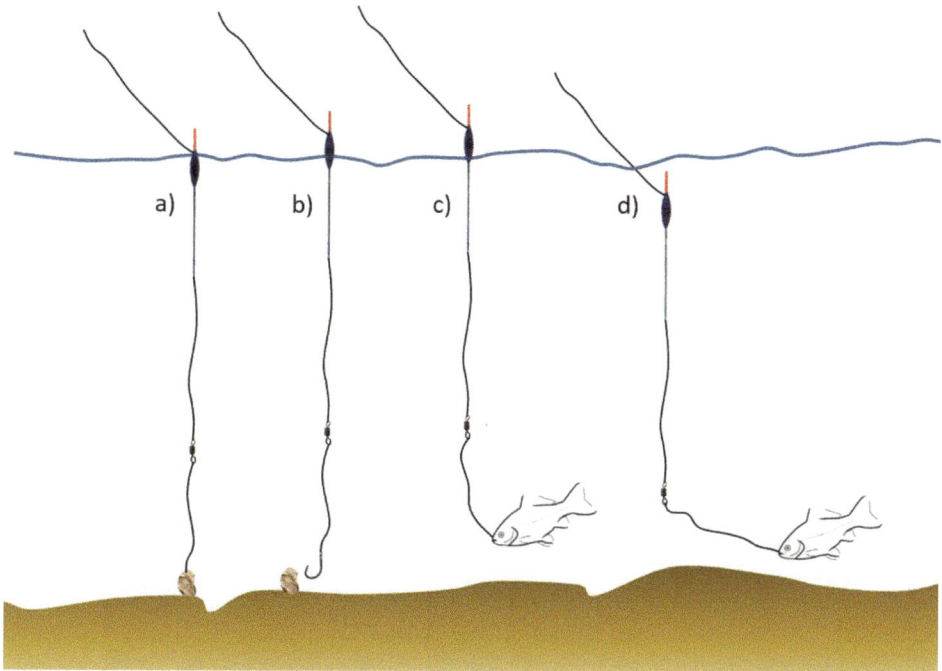

FIGURE 2.8 Dead depth method with fine pole float and bait resting on the bed: (a) float set at dead depth with weight of bait balancing float's buoyancy; (b) float rises if bait falls off; (c) float rises as a fish picks up bait; and (d) float sinks as a fish moves away. (Image © Mark Everard.)

sharp hooks – typically size 16 or 14 – with small baits that are either soft and presented on the hook or on a hair or a hair-rigged bait band. It is important that the feeder or lead runs freely on the line, not so much for bite detection as the self-hooked fish will ensure that bites are visible but to ensure that there is no risk of tethering a fish in the unlikely event of tackle loss.

When tench fishing in particular, it is common to rake a weedy swim to gain access to a corridor of clear water, adding in some loose feed and returning when the swim has been allowed to settle and the tench have come to investigate the cloudier water and the free offerings and natural food that the raking has flushed into the water column. In my direct experience, this approach has not worked for crucians. Crucians seem to eschew the disturbance, and perhaps also feel exposed in opened swims. However, other angling colleagues have found that crucians will respond to raking in the waters they fish. It seems that there is no universally right or wrong way, and that crucian behaviours in different waters may vary. My preference is to fish in undisturbed waters, using watercraft and spotting natural clearings close to cover, or to fish into dusk or the early night when darkness itself is a form of protective cover (Figure 2.9).

Also, bear in mind that silt is a form of cover. Watching the brood stock crucians in my garden pond, often all I see by day is subtle movement in loose bottom sediment.

FIGURE 2.9   A chunky crucian caught by float fishing tactics against a classic Rapidex centrepin reel. (Image © Mark Everard.)

Rather than swimming in clear water, as we assume fishes always do, my crucians actually swim invisibly in the organic matter in the pool bed when they are not secreting themselves amongst the marginal plans. I have yet to devise a ploy to catch crucians that are probably feeding opportunistically under a layer of muck on a pond bed, though arguably the short hook link of a self-hooking leger rig might be achieving this and not, as we imagine, because the bait is wafting in open water above hard lake or pool sediment.

I also sometimes deploy freeline tactics, at least in daylight. This is an approach I used a lot in the 1970s, but for which I rediscovered my passion when I was writing my 2013 book *Redfin Dairies* in which I recounted how it was then my 'go to' method. I have used it many times when crucian fishing, a simple pinch of breadflake on a size 12 or 14 hook tied direct to 2lb nylon monofilament line greased with Vaseline. The line is incrementally pulled down by the slow descent of the almost neutrally buoyant flake, and instantly responds to movements imparted by inquisitive fish. The method still works, including for crucians as much as for roach, rudd, grass carp (with heavier line), chub, dace and many other species. It is also hugely enjoyable, and the direct fight with nothing between angler and fish is a joy.

Whatever the presentation method, subtlety is the key. That includes approaching swims in which crucians may literally be right under your feet. Also, it applies to delicate baiting with free offerings and presentation of the hook bait, bite detection with float as light as you can get away with and dotting it right down to the tip, and playing the fish firmly but carefully on forgiving tackle.

LURE FISHING FOR CRUCIANS

One of the interesting recent trends in angling in Britain over recent years, catching up with innovations in the United States and in continental Europe, has been seen in the wider uptake of lure fishing. This extends well beyond the larger predatory freshwater fishes – pike, perch, zander, warm-water chub, trout and salmon – but deliberately to target many other opportunistic species. Not just bigger species like barbel and carp either. The pursuit of all manner of fishes great and small is enabled by a new generation of micro-lures paired with delicate rods and other kit to make them work optimally.

In reality, virtually all fish are opportunistically predatory, all feeding on small invertebrates as fry for necessary nutritional boost. Many more species are increasingly carnivorous as their metabolic rate rises in summertime, and when a smorgasbord of fish fry as well as larger invertebrates becomes available. Dace, roach, rudd and many other coarse fishes generally thought of as non-predatory become easy prey to the angler fishing fry or invertebrate-imitating flies or micro-lures. This includes smaller fish species too: gudgeon, ruffe, minnows, sticklebacks and many more besides.

The Scandinavian website noted previously documenting fishing for crucians in winter, including through holes in the ice, documents the use of small jigs as well as baited hooks, providing details of jig type and location.[2] But, as described, the fish photographed on the site not only have deeply forked tails but also a suspiciously low number of scales along the lateral line (hard to count definitively due to low image resolution) so appear to be gibel or brown goldfish rather than crucians. That said, the approach is undoubtedly worth pursuing for crucians, using a short rod with a soft tip to present small jigs of natural colours – grey or golden are suggested – on a small hook, jigging the lure up and down slowly near the bottom and next to the cover of vegetation. An angling friend of mine has successfully caught crucians with micro-lure as well as artificial flies, though recommends getting the crucians feeding in a swim first before trying this method.

THE FIGHTING CRUCIAN

As John Bailey wrote of in his 1992 book *Fisherman's Valley: Seasonal Tips for Coarse Anglers*,

> *Their body is not built for long powerful runs but more for short stabbing dives, so typical of a crucian on the end of the line.*

As noted in the previous chapter of this book, I have to introduce water gently into the crucian-containing fish tanks dotted around my house as the slightest turbulence makes the fish tumble. They are ill-adapted to stronger currents, and therefore to

FIGURE 2.10 A healthy crucian lying on wet netting on a classic Abu 501 reel. (Image © Mark Everard.)

sustained runs as experienced, for example, when playing a trout, chub or barbel. This is reflected by the way the crucian resists the angler when hooked (Figure 2.10).

The fight of a crucian is indeed characteristic. The fish is undoubtedly strong, but the description 'short stabbing dives' sums it up nicely. Add to this some of the characteristic 'fluttering' fight of a decent silver bream, both fish sharing a similar high-backed and laterally compressed body profile. Also, as regular crucian anglers already know, these obliging fish have a habit of swimming round in erratic circles when hooked. This can be helpful when they are hooked close to potential snags!

It is vital that the tackle used is forgiving. This includes a suitably soft-actioned rod or whip, or a suitably elasticated pole. A reel, if used, should also have a smooth drag, or in my case a strong preference for a centrepin or else backwinding with a fixed spool reel. It is easy to bounce the hook out of the mouth of a jagging crucian as it seeks sanctuary, and particularly at very close range when the elasticity of the line is minimised.

PHOTOGRAPHING CRUCIANS

So, you have your specimen crucian in your net, or have a bag of them. What next?

From a personal point of view, I would not have a bag of fish. I do not use a keepnet, preferring to put my fish back straight away to avert any possibility of damage. If I need to weigh the fish of any species, or to take a quick photograph, I may hold the fish in

FIGURE 2.11 The author with a 2lb 13oz crucian photographed using a voice-activated shutter release app on a mobile phone. (Image © Mark Everard.)

the landing net head for a minimal time whilst I zero the scales and set up the camera. The fish then comes out of the water in a smooth sequence into a wetted weight sling on a moistened unhooking mat or soft, wet grass, recording the weight before then posing for the photo.

In the modern era, I use my smartphone for the photograph using a voice-activated shutter release (Figure 2.11). The use of a smartphone with voice-activated shutter release – various apps are available for download – is a great advance for fish-care. You can see what the photo will look like before you take it, making sure that the composition is right without cutting off bits of the fish or your head. When everything looks right, you can then trigger the shutter without having to put the fish down on the mat or to wait for a timer on the shutter release, which always seems to coincide with a fish kicking in your hands. You can then see what the photograph looks like before gently releasing the fish back into the water, with minimal disruption. Then, after drying your hands, you can put the kit back into a safe place secure in the knowledge that the fish has been as well cared for as possible.

The exception is when I decide to retain the fish for a longer period – as short as feasible – if I need to find a witness, or if I have a kit malfunction. For this purpose, I keep a live-bait holder or a pike tube in my kit bag. The soft, black mesh of the holder or tube, staked carefully to maximise room and water exchange through the larger pores, means that the fish are kept in as dark and stress-free an environment as possible. This can enable the fish to be held safely and in a tranquil state for a considerable time, possibly until the end of a major rain-storm or into daylight if caught late in a night session.

Fish care is our highest duty. Whilst weighing and photographing fish are vanities for the angler rather than done for the sake of the fish, these pieces of advice can create the least stressful conditions for the fish. We owe the fish that much.

A final note is that the obliging crucian tends to lie still in the folds of a moist net when landed, not leaping around like a grayling or brown trout. This makes life a lot more relaxed for both fish and angler. It also makes photography a lot easier. However, keeping the fish moist and minimising air exposure obviously remain priorities.

CRUCIAN RECORDS IN MODERN TIMES

Angling records within the United Kingdom, Northern Ireland and the Channel Islands (collectively the 'British Records') are the responsibility of the British Record (Rod Caught) Fish Committee (BR(RC)FC). The BR(RC)FC was formed in 1968, and since 2009 has been part of the Angling Trust.

The British Rod-caught Record crucian at the time of writing – possibly one that may be beaten by the time you read this book! – is a fish of 4lb 12oz (1.70 kg) caught on 10 April 2021 by Julian Barnes (Figure 2.12). The fish was originally weighed and reported in the weekly magazine *Angling Times*[3] at 4lb 14oz, but it was subsequently ratified at 4lb 12oz after an adjustment when the scales were tested by Norfolk Trading Standards and found to be weighing one-and-three-fifths ounces over. This mighty crucian was captured from Johnson's Lake at the Godalming Angling Club's Marsh Farm complex in Milford, Surrey. This record was ratified by the BR(RC)FC on 19 July 2021, in a remote online meeting necessitated by Covid restrictions.

Julian Barnes, the jubilant captor who had made the historic trip to Johnson's Lake from his home in St Neots, Cambridgeshire, was reported in the *Angling Times* as stating:

> *Come dawn, I introduced five Spombs full of hemp and caster to a small feature 40 yards out from the bank. Over the top of this I decided to use method feeder tactics, with 35g inline feeders carrying my special groundbait mix, and short hook-links with fake caster hookbaits. I recast the rods every hour to keep the swim topped up and, hopefully, the fish feeding."* After banking crucians weighing 4lb 6oz and 4lb 3oz and a 6lb 5oz tench on the first day of the session, he decided to stay for another day and *"At dawn the next day I baited up again, and at 10.30am I had a bite from what turned out to be this most incredible fish – a new British Record crucian!"*

The previous British Record crucian was held by a fish of 4lb 10oz caught by Craig Smithson on 3 August 2020 from Milton Abbas Fishery (Milton Abbas, Dorset). Craig's crucian was initially reported as weighing 4lb 11oz, though this was rounded down after the scales were tested. An article in the weekly magazine *Angling Times* tells the story of the capture of this remarkable fish.[4] The crucian was caught when Craig Smithson was fishing for common carp during a 48-hour stay on the eight-acre

FIGURE 2.12   Julian Barnes holding his stunning 4lb 12oz British Record crucian. (Image © Julian Barnes.)

lake Milton Abbas syndicate lake in Dorset. The fish was taken on a red 15 millimetre fishmeal pop-up boilie positioned next to a pre-baited area adjacent to a lily bed. The fishery owner was summoned, joined by three other syndicate members, to witness the weighing and measuring and to take photographs. A claim was submitted to the BR(RC)FC, along with a set of scales used to weigh the crucian, a list of the names and contact details of the witnesses and two scale samples to verify that the fish was not a hybrid. Crucians had been stocked into the lake roughly 20 years previously. Some of the few remaining, larger crucians had been caught by another angler and also in a routine netting.

Prior to Craig Smithson's fish, two crucians each of 4lb 10oz were caught by Michael James on 4 May 2015 and Steve Frapwell on 10 May 2015, both fish taken from Johnson's Lake in Godalming, Surrey, the renowned large crucian water.

Previous to these fish, British crucian record-holders included fish of:

- 4lb 9oz, caught by Peter Cardozo (2 May 2015, Johnsons Lake)
- 4lb 9oz, caught by Joshua Blavins (17 August 2011, Moor Mill Gravel Pits, Frogmore, Hertfordshire)

- 4lb 9oz, caught by Philip Smith (7 August 2004, RMC Summer pit, Yateley, Surrey)
- 4lb 9oz, caught by Martin Bowler (16 May 2003, Little Moulsham Lake, Yateley)
- 4lb 8oz, caught by Jay Allen (2000, RMC Summer pit, Yateley)

CRUCIAN RECORDS IN FORMER TIMES

Alwyne Wheeler documents in his 1969 book *The Fishes of the British Isles and North West Europe* that

The British record fish weighed 4 lb 11 oz (2.12 kg).

Wheeler refers to the same fish documented by Eric Marshall-Hardy in his delightful little 1943 book *Coarse Fish* written during the Second World War:

Writing in the Fishing Gazette of November 29th, 1941, Mr. Wm. Berry cites a Crucian Carp taken by Mr. H. C. Hinson, weighing 4 lb. 11 oz. This aroused my curiosity, as if it were authenticated it must take its place as the British rod-caught record. I got into communication with Mr. Hinson, who has satisfied me completely regarding this fish. These are the facts. On September 25th, 1938, Mr. H. C. Hinson took part with some sixty other anglers in a match at Broadwater Lake, Godalming, Surrey. On that occasion there were only two fish taken—one of 13 oz. and Mr. Hinson's 4 lb. II oz. Crucian Carp. His fish was " weighed in " by a member of the Godalming A.C.—Mr. A. Johnson, who is at the time of writing serving in the East. Mr. Hinson writes: " It is so long ago and so much has happened since that even if I could recall them, no other member who was present at the 'weigh-in 'would commit himself." However, Mr. Hinson entered his catch in a newspaper competition on the following week, not being aware that it was of record weight. The fact would appear to have escaped the judge's attention also, for the entry was not acclaimed as a record. As for me, and this book, I am satisfied to hold Mr. Hinson's fish as a new British record, which wrests the title from the splendid fish caught by Mr. E. Palmer from Bradfield Combust in 1933, which weighed 4 lb.

However, the BR(RC)FC took a controversial decision shortly after its formation to purge the pre-existing British Record list of former records that could not be verified with photographic evidence, witnesses, tested weighing scales, correct species identification and other supporting evidence. One of the many casualties of this purge was the former British record crucian weighing in at 4lb 11oz. This long-standing record fish was the one captured by Mr H.C. Hinson on 25 September 1938 from Broadwater Lake, Godalming, Surrey. The record was not officially broken for another 62 years.

So how big can crucians grow? Eric Marshall-Hardy also records

It will be seen that Crucians are comparatively diminutive. Continental claims for the weight of Crucian Carp do not exceed 7 lb. so far as I can ascertain. Small as they are, they are hardy fishes, which can be transported in damp moss for great distances.

There is plenty of evidence of fish growing to 6lb and even 7lb in Scandinavia. In his wonderful 2010 book *Crock of Gold: Seeking the Crucian Carp*, Peter Rolfe records that

> *In 2008 or 2009 a crucian carp 44cm long was caught in a fish trap in Northern Sweden, reported as weighting 7lb 5oz. I've seen the photograph, though it was not of good enough quality to reproduce here, and I have no doubt of the fish's genuineness.*

Size though is not the ultimate purpose of fishing and certainly not the most satisfying element. Exploring wild and wonderful places, socialising with like-minded people and making the best of the waters to which you have access is challenge enough. My spring 2024 mission was to catch a crucian over 2lb from waters not so far from my home, and particularly from places where they were either not known or else were rarely reported. At times frustrating though always enjoyable, I enjoyed many blank sessions but also periodically had some superb catches of prime crucians including a new personal best specimen (Figure 2.13): A real old 'warrior' of a crucian of 2lb 14oz caught on two maggots on a size 14 hook presented with a pole virtually at my feet in a lake margin!

FIGURE 2.13   The author holds his 'old warrior' personal best crucian of 2lb 14oz crucian after a satisfying spring 2024 mission on local waters. (Image © Mark Everard.)

CRUCIAN MATCH CATCHES

Whilst crucians, or more often a mix of *Carassius* and their hybrids of less clear prevenance, are stocked into commercial fisheries that are match-fished, there are no records of record crucian match catches (Figure 2.14).

FIGURE 2.14   A fine net of crucians. (Image © Mark Everard.)

CRUCIANS AS BAIT

Crucians are small and robust fish and, as we have seen, appear to be favoured as prey by pike, perch and other predatory fishes. For these reasons, they appear to be ideal live baits when fishing for predatory fishes. In fact, crucians are widely used as live baits for pike in Sweden, where these fish are far more common.

However, we have to bear in mind wider conservation issues (quite apart from my personal dislike of using live fish as bait). These wider considerations relate to the conservation of residual populations of this now-less-widely distributed fish. Also, issues of confusion between closely related species and their hybrids – gibel, goldfish, nigorobuna and common carp in particular – and the potential for inadvertently spreading them into virgin waters or pre-existing crucian waters. The best advice is simply to choose other species as live baits or use dead baits.

CRUCIANS AS PROBLEMS

For all the virtues of the crucian, there remain conservation concerns in some of the places where they have been introduced beyond their native range. The Food and Agriculture Organization of the United Nations lists crucians as potential pests.[5] Nonetheless, there is little or no evidence of instances of crucians creating problems as invasive fishes despite their broader introductions.

Crucians have been found in Australia, presumed to have been released from aquaculture, and are listed as of potential concern though with no evidence of actual impacts.[6]

There also remain concerns about crucians in the United States, despite no evidence that they are found in the wild. The US Geological Society website[7] at the time of writing notes that

> There are no recent reports of crucian carp in the U.S. An earlier report that either the crucian carp or a hybrid (with goldfish) had been introduced into Texas... is now considered unlikely. The introduction and status of this species remains uncertain...
>
> The impacts of this species are currently unknown, as no studies have been done to determine how it has affected ecosystems in the invaded range. The absence of data does not equate to lack of effects. It does, however, mean that research is required to evaluate effects before conclusions can be made.

THE POST-CARP ENVIRONMENT

Today, common carp, once scattered and of almost mystical status, seem to be stocked or overstocked in every available hole in the ground. They certainly dominate the economics of modern angling in the United Kingdom. And this notwithstanding the damage that common carp inflict upon the waters into which they are stocked, both here and right around the world.

However, there is a groundswell of interest in angling in specialist crucian and tench waters, not just amongst older and perhaps wistful people like me but also amongst growing numbers of younger anglers jaded by the automation and single dimension of the modern carp scene.

Prime amongst these specialised crucian waters is the Marsh Farm complex at Enton, near Godalming, Surrey, owned and run by the Godalming Angling Association. This remarkable crucian mecca had started out life as a single lake purchased in 1985 – Enton Lake, now known as Johnson's Lake in honour of a now sadly deceased club stalwart – but has since been augmented substantially after the Godalming Angling Association purchased an adjacent chicken farm in 1997 on which they built three lakes that opened for angling in 2004.

The Marsh Farm complex though is not alone with a rising number of more carp-free or low-carp stillwaters managed for traditional fishing, particularly for crucians and tench. Quality assurance of the heritage of these crucians is backed up by DNA testing. Portsmouth & District Angling Society is at the forefront, founded in 1948 and now one of the leading angling clubs in the South of England with five-year plans for each of its multiple venues offering a variety of different types of fishing to suit all angling inter-ests. One of the Portsmouth club waters is Abshot Pond, near Fareham in Hampshire – a classic shallow farm pond with ten swims all with features such as lilies or reeds and managed to support a large head of 'true' (DNA-verified) crucians as well as rudd, roach, perch and specimen tench. Abshot is categorised as a Class A 'Managed as Premium Crucian Fishery' under the NCCP classification, explained in Chapter 3 of this book.

This post-carp scenario, in many ways, harks back to former times when common carp were seemingly not in every fishery, a golden age perhaps albeit that the numbers of smaller water bodies in this modern world are far fewer.

RUNNING A CRUCIAN FISHERY

I will touch only lightly on the topic of running a crucian fishery, and for two very good reasons. Firstly, I have never been responsible for one, though I am privileged to fish one run by a friend, and I also host a breeding crucian population in two of my garden ponds. Secondly, I could not improve on what Peter Rolfe has written on the topic in his wonderful 2010 book *Crock of Gold: Seeking the Crucian Carp* and its 2024 revision *Crock of Gold: The Crucian Revealed*. These volumes synthesise Peter's then 50-plus years of hands-on experience. Peter has added to this in various other publications, both printed and online (see www.crucians.org).

The work of the National Crucian Conservation Project (NCCP), addressed in the following chapter of this book, augments our knowledge still further. If there is one piece of advice I would pass on from Peter, it is that the ideal crucian fishery has at least two separate ponds.

- One is a stock pond, fertile and well-weeded, in which crucians can breed and recruit freely. This breeding pond may need to be netted, possibly after draining down for which design considerations need to be planned from the outset. This enables stock to be thinned, removing excess crucians for sale or onward stocking. Predators should be avoided: certainly pike, but also perch have a taste for any mouth-sized crucians and can curtail their numbers. Roach and rudd are also best avoided as they too can breed prolifically, potentially forming stunted populations and suppressing food availability for the crucians. Tench, however, seem to be an ideal pool-mate, also well-adapted to small pond environments and not overly competitive with crucians. The breeding pond in my back garden is of this type, small but productive from which I make small crucians available to others the following spring once they have grown on.
- The other type of pond is a mixed fishery into which crucians are stocked, not in any expectation that they might breed but where they can grow larger in the absence of stiff competition from hordes of their own kind. In some ponds, crucians can do well in low numbers, growing on to often specimen size. In others, they simply fail to establish. But it is in these larger, mixed bodies of water, particularly those managed to maintain good habitat diver-sity with a low predator density, that giants can grow perhaps undetected.

This 'two pond solution' does seem to hark back to the ways that crucians seemed to have been freely moved informally between ponds when both ponds and fish were far more common across rural landscapes. In those far-gone days, some pools were known as 'mother ponds', prolific in crucians and from which fish could be caught and moved – often in buckets suspended over the handlebars of bikes – to restock adjacent ponds.

One of the many changes that have occurred in the intervening period is the level of avian predation encountered in a much-altered British landscape, including networks of gravel pits and other open water bodies that have set out a welcome mat for cormorants to move inland. Cormorants are efficient hunters of fish and, in a confined space, can decimate stocks. The poor crucian seems to be a particularly attractive 'bite-sized snack' for many predators, including the voracious cormorant. A high density of habitat features – weed beds, reed stands, fallen trees, man-made gabions, submerged netting of a large enough mesh size not to impede the passage of fish – can certainly interfere with the underwater hunting efficiency of cormorants, but will not entirely prevent their impacts on smaller water bodies. Some vigilance will be required to deter these birds that, though only finding food as they are evolved to do, can decimate a small fishery.

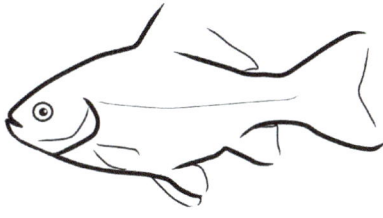

NOTES

1. Wakeowl.com. (Not dated). *Winter fishing: we catch crucian carp with a jig.* [Online.] https://www.wakeowl.com/fishing-crafts-for-the-fisherman/winter-fishing-we-catch-crucian-carp-with-a-jig/, accessed 29 December 2023.
2. Wakeowl.com. (Not dated). *Winter fishing: we catch crucian carp with a jig.* [Online.] https://www.wakeowl.com/fishing-crafts-for-the-fisherman/winter-fishing-we-catch-crucian-carp-with-a-jig/, accessed 29 December 2023.
3. Angling Times. (2021). British Crucian Record Smashed! *Angling Times,* 4 May 2021.
4. Angling Times. (2020). British record crucian carp – the full story. *Angling Times,* 06 August 2020. [Online.] https://www.anglingtimes.co.uk/news/stories/british-record-crucian-carp-the-full-story/, accessed 27 December 2022.
5. FAO. (1997). *FAO Database on Introduced Aquatic Species.* Food and Agriculture Organization of the United Nations (FAO). [Online.] http://www.fao.org/figis/servlet/static?dom=collection&xml=dias_collection12.xml&xp_detail=med, accessed 29 December 2022.
6. Bray, D.I. (2022). *Carassius carassius* in Fishes of Australia. Museums Victoria. [Online.] https://fishesofaustralia.net.au/home/species/2693, accessed 29 Dec 2022.
7. USGS. (2022). NAS – Nonindigenous Aquatic Species. US Geological Survey (USGS). [Online.] Crucian carp (*Carassius carassius*) – Species Profile (usgs.gov), accessed 29 December 2022.

Crucians and People

3

What is it about the crucian that so endears them? Partly, I think, it is their familiarity to many of us who enjoyed tracking them down amongst the wealth of small farm pools in southern and eastern England in the 1960s and 1970s. Many of those ponds have since been swept aside with changes in land use and also by the pervasion of introduced common carp that were once rarely found but are now hard to avoid. It is so good to see a resurgence in interest in crucians (Figure 3.1), and the conservation of both the species and the diminishing habitats that support them.

But crucians are also the focus of a range of social values. Let's explore some of these in this chapter.

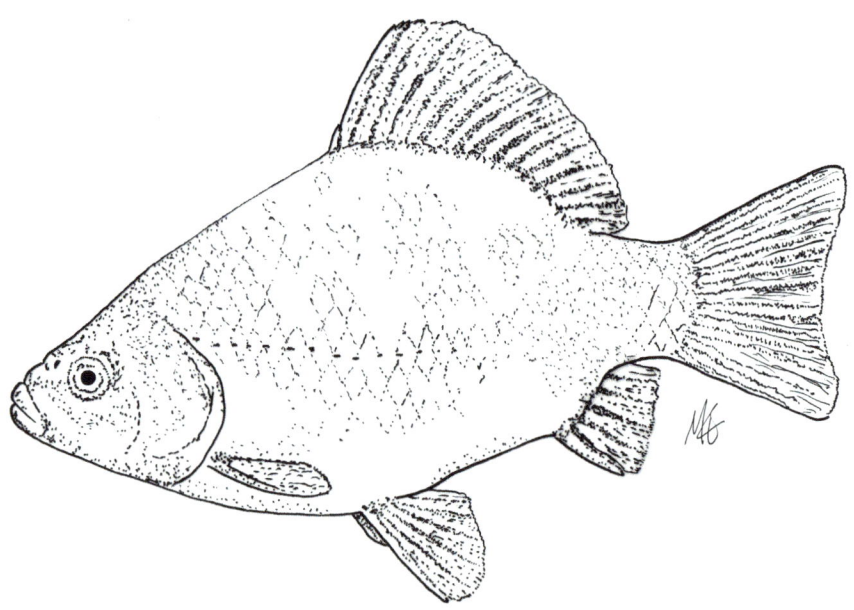

FIGURE 3.1 Crucians have different meanings to different people. (Image © Mark Everard.)

DOI: 10.1201/9781003560791-3

CRUCIAN ETYMOLOGY

The common name of the crucian was derived in around the eighteenth century from the Low German *Karusse*.

As discussed earlier in this book, crucians come under the genus *Carassius*, not *Cyprinus*, so calling them 'crucian carp' can be misleading as these are quite different fish from their bulky cousins of the genus *Cyprinus*. We call goldfish simply 'goldfish', and under the same convention let's call gibel simply 'gibel'. So, 'crucian' it is!

As we have seen in previous chapters, various other localised common names have been applied in times gone by: 'bream carp', 'Hamburgh carp' and 'German carp' amongst them. A local Norfolk name for the crucian was the 'mudfish', named for its capacity to survive in mere puddles until rains finally came to replenish drought-stricken ponds.

Crucians have a wide natural European and Western Asian range and have also been introduced to many countries. They therefore have a broad variety of common names. In fact, perhaps owing to their scattered distribution in small ponds and other fragmented habitats, they have a very wide range of common names even within some of the countries in which they occur. Some of these common names are included in Table 3.1.

Whilst the scientific name *Carassius carassius* (Linnaeus 1758) has long been accepted in science, breaking this name down into its component parts tells us something more about this fish.

The *Carassius* element of the name is self-explanatory, with an initial capital 'C' for the genus and an initial lower case 'c' for the specific name. The extension to the Latin name is known as the 'authority', naming the person who first described it scientifically and the date it was described. The extension to the Latin name of the crucian, (Linnaeus 1758), tells us that Carl Linnaeus was the first to describe this species in 1758. However, the parentheses (brackets) around the author's name and the date tell us that the species was originally described under a different name. The original Latin name that Carl Linnaeus bestowed on the fish was *Cyprinus carassius* Linnaeus, 1758. In fact, Linnaeus initially classified a lot of carp-like fishes under the wide genus *Cyprinus* (Figure 3.2).

Many authors throughout history have assigned Latin names to fish they considered different species, but that are now known to be crucians. Many of these diverse but now-defunct Latin names are listed in Table 3.2.

A further Latin name, *Carassius gibelio* (non Bloch, 1782) is misapplied, referring to the genetically close gibel.

OTHER MEANINGS OF THE WORD 'Crucian'

Aside from the common name of *C. carassius*, the word 'crucian' describes a number of different things.

TABLE 3.1 Common names by which crucians are known across the world

LANGUAGE	COMMON NAMES
Amharic (Ethiopia)	Bilcha
Czech	Karas bahenní, Karas obecný, Karas obycajny
Danish	Karuds, Karudse, Karusse, Søkaruds, Søkarusse
Dutch	Kroeskarper, Steenkarper
English	Crucian, Crucian carp (also in different parts of the US English carp, Gibele, Golden carp)
Estonian	Harilik koger, Koger
Finnish	Ruutana
French	Carache, Carassin, Carassin commun, Carouche, Carpe à la lune, Carreau, Cyprin, Cyprin, Gardon carpé, Gibèle, Meule
German	Bauernkarpfen, Boretsch, Breitling, Burretschel, Gareisel, Gareisel, Gareisl, Garrausche, Garusse, Geibel, Giebel, Goldkarausche, Goldkarpfen, Guratsch, Gurretfisch, Halbgareis, Hungerkarausche, Kanas, Karatsche, Karausche, Karausche, Karausse, Karras, Karusche, Karuske, Karutze, Kotbuckel, Kotkarpfe, Kotplette, Kotscheberl, Krus, Krutsch, Kuretschel, Mölenke, Moorkarpfen, Moorkarpfen, Schlammkarausche, Schneiderkarpfen, Seekarausche, Steinkarpel, Strummer, Sumpfkarausche
Greek	Koutsouras, Petaloúda, Κουτσουράς, Πεταλούδα, Τεκέδι
Hungarian	Kárász
Italian	Carassio
Japanese	Funa
Kashmiri	Gang gad
Latvian	Karūsa
Lithuanian	Auksinis karosas
Nepali	Rato machha
Norwegian	Karuss
Polish	Karas, Karaś, Karas pospolity
Spanish	Carpín
Polish	Karas, Karaś, Karas pospolity
Portuguese	Pimpao, Pimpão-comum
Rumanian	Caracuda, Caracuda rotunda, Caras, Carasita, Garasita
Russian	Karas' obyknovennyi, Sobo, Карась золотой
Serbian	Karaš
Slovak	Karas obycajný
Slovene	Koreselj
Swedish	Ruda
Turkish	Havuz baligi, Havuz balığı, Kirmizi balik
Ukrainian	Karas zvychajnyi, Карась звичайний, Карась золотий
Albanian	Karasi

FIGURE 3.2   Mixed species from the carp family: (1) fully scaled common carp, (2) mirror carp, (3) crucian and (4) barbel (1884 woodcut).

TABLE 3.2   Former but now abandoned names by which crucians were known

Carassius moles Nordmann, 1840
Cyprinus Carassius Linnaeus, 1758
Carassius carassius Carassius (Linnaeus, 1758)
Cyprinus moles Agassiz, 1835
Carassius humilis Heckel, 1837
Cyprinus charax Lesniewski, 1837
Carassius charax (Lesniewski, 1837)
Carassius vulgaris Nordmann, 1840
Cyprinus moles Selys-Longchamps, 1842
Cyprinus moles Valenciennes, 1842
Carassius linnaei Bonaparte, 1845
Cyprinus gibelio minutus Kessler, 1856
Carassius gibelio minutus (Kessler, 1856)

One of these meanings is a person from St Croix island in the Caribbean Sea, a county and constituent district of the United States Virgin Islands (USVI). It also relates to a dialect of Virgin Islands Creole spoken on St Croix.

The name 'Karaś' (Polish for 'crucian') is also assigned to the Polish 1.50 Złoty 1971 – PZL 23-A 'Karaś' light bomber and reconnaissance aircraft. This aircraft was designed in the mid-1930s by PZL in Warsaw, serving as the primary Polish reconnaissance bomber in use during the invasion of Poland. This bomber featured on a Polish postage stamp in 1973 commemorating Polish war planes in 1939 and the Polish Air Force Emblem. Some 3 million of these multicoloured stamps were printed.

'Crucian' is also a fictional creature in the 'Dungeons & Dragons', a fantasy role-playing game. I have no experience of the game, but who am I to argue with what the internet tells me?

CRUCIAN CUISINE

A few years ago, when I was doing some work involving crucians, a fellow scientist (now sadly no longer with us and who had best anyhow remain nameless) told me that there was no way that crucians could have been transported for food. He had caught and cooked virtually every species of British freshwater fish but, no matter what he did to them, crucians tasted awful (Figure 3.3)!

The Reverend W Houghton praised the flesh of the crucian at best faintly in his 1879 book *British Fresh-Water Fishes*,

> *...and although its flesh cannot be considered dainty food, yet good well-nourished specimens are by no means to be despised.*

There is a Chinese crucian soup recipe,[1] the referenced source noting that

> *This Chinese soup is made with Cilantro and Crucian Carp, a cheap fish found in markets and common in every household.*

Another Chinese recipe is the Beer Crucian Carp.[2]

Questions have, though, to be raised about whether the 'crucians' in question in these Chinese recipes are *C. carassius*, given that other *Carassius* species but not crucians are native to China.

Dried salted crucians though are popular beer snacks in some parts of continental Europe. Perhaps the beer helps make this fish more digestible? As noted in the first chapter of this book, perhaps the drying of crucians to preserve them as a mobile food could account for the remains of a crucian found in a midden dated back to Roman Britain.

The crucian is treated more as a delicacy in Poland where the *karaś* is considered the best-tasting pan fish, traditionally served with sour cream in the dish *karasie w śmietanie*.[3] In Russia, this fish is known as Золотой карась ('golden crucian') and

FIGURE 3.3   A variety of recipes are available for the crucian. (Image © Mark Everard.)

is one of the fish used in a borscht recipe called *borshch s karasei*.⁴ Another classic
Russian recipe is fried crucians in sour cream.⁵

CRUCIANS IN AQUACULTURE

Bent J Muus and Preben Dahlstrom record in their 1967 book *Collins Guide to the
Freshwater Fishes of Britain and Europe* that

> *The crucian carp is sometimes stocked as a pond-fish in central Europe, especially
> in ponds where the common carp will not thrive. A well-developed, deep-bodied
> German race 'Spechthausen' is especially fast growing and matures at 2 years at
> a length of 13–15 cm. The presence of crucian carp is undesirable in carp-ponds
> because they complete for food with the more valuable carp, and partly because they
> form a 'reservoir' for parasites which affect carp.*

Bent J Muus and Preben Dahlstrom also record that crucian hybrids with common
carp were commonly used in aquaculture, noting that

> *They are used in south-eastern Europe partly in fish-farms and partly for stock-
> ing small waters. These hybrids have several advantages, in that being sterile they*

will not overfill the water with stunted fish, whilst they also have the hardiness and resistance to disease of the crucian carp, and a fairly good growth rate inherited from the carp.

The United Nations Food and Agriculture Organization (FAO)[6] reports that

The culture of crucian carp was initiated in China. The earliest activity can be traced back to the East Han Dynasty (25–189 A.D.), according to archaeological studies, to the Song Dynasty (960–1279 A.D.) and according to written records. However, production was limited to a rather small scale. Aquaculture of this species was limited to China and Japan until the mid-1960s. Since then it has gradually expanded to many other countries and regions, including Taiwan Province of China, Belarus, Republic of Korea, and Uzbekistan. The major producer has always been China, whose production has expanded from less than 2000 tonnes in 1950 to nearly 1.7 million tonnes in 2002 (99.6 percent of the global total).

This FAO record though is almost certainly flawed as crucians are not native to either China or Japan, though other *Carassius* species are present. FAO literature also documents that crucians were ranked 9th worldwide in 2008 in terms of production in aquaculture, though these statistics are also flawed as the FAO incorrectly treated Asian gibel (*C. gibelio*) as a subspecies of the crucian (*C. carassius*) and the vast bulk of the recorded production is from China.

CRUCIANS IN LITERATURE

A number of angling and ichthyological titles listed in this book address the crucian (Figure 3.4). However, many more do not.

Conspicuous absentees include none other than Izaak Walton in his 1653 book *The Compleat Angler*. Walton most likely was unaware of this smaller member of the carp clan, perhaps even regarding this smaller fish as a juvenile of the common carp. Or perhaps the recent introduction of this fish to Britain, somewhere between 300 and 500 years ago, meant that it was either still absent or had only a highly restricted distribution remote from waters familiar to Walton and his co-contributors.

Many older anglers, and more than a few book collectors, will recall fondly the *How to Catch Them* series of small pocket books introduced in 1954 by the publisher Herbert Jenkins. They are sometimes referred to affectionately as the 'HTCT series'. These books were hugely popular and enjoyable, uniform in appearance and size. The 44 titles published between 1954 and 1969 introduced specific fish species, techniques or equipment, each written by a well-known angling expert of that era. However, the crucian never featured in the HTCT series – a sad omission reflecting the lower interest in that fish at the time. I hope that this book goes at least some way to filling that niche!

The crucian has, though, received renewed dedicated attention in recent years. This includes not only as an angling target but also in terms of dedicated books. Aside from

FIGURE 3.4   1897 woodcut of a crucian.

the book in your hands, two rather excellent books (noting also one update) dedicated to the crucian are:

- *Crock of Gold: Seeking the Crucian Carp*, a rather wonderful little book – believed to be the first one dedicated to this fish – written by Peter Rolfe and published in 2010 by MPress (Media) Ltd. Peter's book takes a full 360-degree view of all things relating to crucians: as a threatened species, as an angling target, how the literature has addressed crucians over the past 300 years, issues of identification and hybridisation, and insights on creating and managing crucian waters based on his long-term, hands-on experience.
  - As a footnote, Peter had hoped to update the book in 2023 but the publisher did not perceive a market despite substantial interest in, and high market prices associated with, second-hand copies of the original. Consequently, Peter included a final chapter in his 2023 book *Old Angler Rambling* noting further novel understandings and developments since *Crock of Gold* was published. Talking with Peter, I decided that my own crucian book would be a useful addition to the crucian library. It came as a bit of a surprise that Peter then updated his book as the 2024 *Crock of Gold: The Crucian Revealed*!

- *Willow Pitch VI: Crucian Renaissance*, a multi-authored edition compiled and edited by artist and author Chris Turnbull, was published in 2021 by Little Egret Press/Water's Edge Publications. The book contains original contributions from a wide range of well-known crucian aficionados. Regrettably, I declined an invitation to contribute due in part to pressures of work and also because I had not caught a big crucian in many years and felt ill-equipped. This decision was silly really, as there is so much more to be said about this charming little fish than capture of a specimen, as I hope is evident from the contents of these pages.

CRUCIANS AND THE CREATIVE ARTS

Aside from the books already discussed dealing with crucians, there are no creative books dedicated to the crucian in the ways, for example, that Henry Williamson immortalised the Atlantic salmon in his *Salar the Salmon*. Nor has any composer been inspired by the crucian in the way that Franz Schubert composed 'Die Forelle' ('the trout'). Furthermore, crucians have evidently not inspired heraldry, symbology, myth and legend.

Crucians though have featured in some notable paintings. Amongst the best-known is a painting by Alexander Francis Lydon (1836–1917; Figure 3.5) featuring the 'crucian carp' alongside the 'Prussian carp'. Lydon was a British watercolour artist, illustrator

CRUCIAN CARP   PRUSSIAN CARP

FIGURE 3.5   Painting of crucian and Prussian carp by Alexander Francis Lydon (1836–1917).

FIGURE 3.6   Painting of a crucian by Marcus-Eliezer Bloch.

and engraver of natural history and landscapes. Lydon's painting is one of a set illustrating The Reverend W Houghton's 1879 book *British Fresh-Water Fishes*. This series of paintings has since been widely used in tableware and many other products.

Another attractive piece of art concerning the crucian, if lesser known in Britain, is one of a series of prints of fish by Marcus-Eliezer Bloch (Figure 3.6). Bloch, born in 1723 in Bavaria, studied natural history and medicine in Berlin and Frankfurt, where he practised as a physician until his death in 1799. Fish, and a detailed knowledge of ichthyology, remained an abiding passion throughout Bloch's life. He was to publish the 12-volume *Allgemeine Naturgeschichte der Fische* ('General Natural History of Fish') between 1782 and 1795.

More familiar in Britain is a painting of a crucian, artist not identified, that featured as No.4 in a series of 50 tea cards by the Brooke Bond company in the 1960s (Figure 3.7). Though anglers and others with interests in fish of a 'certain age' will doubtless instantly recognise this image, no doubt with fond memories, the series of paintings has been very widely reproduced and used for other purposes ever since, so may be familiar to many more people.

Many British anglers must also have seen, and probably bought, an Environment Agency 2015–2016 non-migratory trout and coarse fish angling licence featuring one of David Miller's excellent paintings of a crucian (Figure 3.8).

Another painting of a crucian seen by many in the United Kingdom is on a series of six First Class postage stamps (Figure 3.9; also including perch, European eel, caddis fly larva, Arctic charr and common toad) issued in June 2013. These stamps were on sale for 60 pence.

Other countries too have featured the crucian on their stamps, including Bulgaria, Ukraine, Poland and the State of Oman.

FIGURE 3.7   Painting of a crucian featuring as No.4 in a series of 50 tea cards by the Brooke Bond company in the 1960s.

FIGURE 3.8   Painting of a crucian by wildlife artist David Miller, featuring on a 2015-2016 Environment Agency fishing licence. (With permission from David Miller.)

FIGURE 3.9   Painting of a crucian featuring on a British postage stamp, June 2013. (With permission from Post Office.)

A range of out-of-copyright woodcuts and other drawings of crucians also feature throughout this book, along with various of my drawings of crucians, baits and fishing tackle.

CRUCIANS AND NATURE CONSERVATION

Freshwater fishes play important roles as key functional elements of aquatic environments, maintaining the integrity of whole ecosystems and the many ways in which they support human needs. Conservation of nature is therefore far more significant than an altruistic concern, as humanity and nature are fully interconnected and mutually interdependent. This awareness is still slowly dawning on society as we face a frightening biodiversity crisis with serious, if massively underappreciated, implications for our future security.

Changing Thinking About Approaches to Nature Conservation

Emerging realisation of the depth and multiple dimensions of fully interlinked environmental and societal crises raises interesting questions about the objectives of nature conservation initiatives. This is particularly so in the light of increasing difficulties inherent in re-creating past ecosystems, which may in practice often be impossible in the light of substantial climatic shifts, massive landscape changes and today's substantially elevated human population that, globally, has tripled in my lifetime alone and is still rising. These large changes in environmental conditions may mean that some nature reserves, pools and other habitats that were once havens for particular species and assemblages of organisms can no longer provide the conditions required for them to prosper, or even to survive. Yesterday's so-called 'fortress conservation' model, founded on preserving conserved areas in as 'natural' a condition as possible, may not only no longer work, but may potentially even hasten the demise of organisms restricted into places of changing character from which they cannot move through surrounding inhospitable environments.

Recognition of inherent dangers in the 'fortress conservation model', albeit that it is still a vital tool for preventing the extinction of the most endangered species, has been driving new ways of thinking and approaches to nature conservation since early in the twenty-first century. Some were summed up by the UK's 2011 'Lawton Review'[7] of nature conservation, which called for a "bigger, better and more joined up" approach. 'More joined up' was a critical element, recognising the need for species and ecosystems to become more mobile across less hostile landscapes.

Further evolution in nature conservation thinking is refocusing on ecosystem complexity and functioning. This shift in approach recognises the need to work with natural processes as a foundation for ecosystem restoration, rather than to seek simply to replace expected species. A pivot to a functional approach represents a more progressive means for addressing the pressing biodiversity and climate crises that humanity faces today in the light of radical changes we have inflicted on habitats and ecosystems. The paradigmatic change in emphasis in conservation thinking includes addressing the roots of ongoing degradation of the many beneficial natural processes that ecosystems provide, ranging from the cleansing of water, air and soils through to capacities for production of food as well as recreational and economic resources.

This is a paradigmatic change that is highly germane to fisheries and the aquatic ecosystems that support them. In fact, freshwater ecosystems are amongst the most threatened at a global scale. They are particularly vulnerable because they integrate pressures from land use, industry, infrastructure and domestic activities across whole, wide catchment landscapes. The world's biodiversity crisis is impacting fresh waters more acutely than other ecosystem types, with around 35% of wetlands globally having been lost in the past 50 years. Freshwater fishes are therefore understandably amongst the most vulnerable groups of organisms globally. 51% of all fish species live in fresh water – over 18,000 species with more discovered each year – making up a quarter of total global vertebrate species.[8] Migratory species are particularly vulnerable due to factors such as increasing hydropower, overfishing, climate change and pollution, with an average global collapse in populations of 76% since 1970 including a catastrophic 93% collapse in Europe.[9]

Shifting Approaches to Nature Conservation and the Crucian

Whilst crucians are distinctively not a migratory fish, the loss of their habitats, as well as invasion by gibel, goldfish and common carp as well now of nigorobuna, means that they may be just as vulnerable to extinction, or perhaps even more so given threats to the smaller water bodies to which they are adapted (Figure 3.10). Freshwater ecosystems are in trouble worldwide, but the fate of smaller waterbodies is particularly overlooked with the bulk of attention diverted to the plight of larger river systems. Yet ponds and wetlands comprise around 80% of the global freshwater environment.[10] In the United Kingdom, they are known to be declining in quality, the Countryside Survey of ponds across Great Britain documenting that two-thirds of high-quality ponds have lost plant species over a 24-year period.[11] Also, unlike rivers and streams, these waterbodies are not routinely monitored or systematically protected. Paradoxically, small waterbodies include many of the best-quality remaining freshwaters, with ponds found to support greater biodiversity than larger waterbodies such as rivers or lakes. They are therefore important targets for protection in the face of the deepening freshwater biodiversity crisis.

These observations and assumptions about the vulnerability of nature, and particularly that found in small water bodies, is consistent with a 2021 assessment of the UK's wildlife, which concluded that there was an average 19% decline in abundance of terrestrial and freshwater species, a 13% decrease in distribution of invertebrates and a 54% decrease in distribution of flowering plants since 1970,[12] though this assessment did not explicitly address freshwater fishes.

FIGURE 3.10 Gibel are widespread across continental Europe, presenting a serious conservation challenge to the crucian. (Image © Mark Everard.)

Finding societally compelling reasons to protect dwindling pond and wetland habitats is therefore an urgent priority. Though crucians may not be strictly native to Britain, they are non-invasive and are characteristic of threatened habitat types. They also have charismatic and heritage attributes. Benign to other wildlife adapted to the vulnerable and significantly reduced resource of pond habitats, crucians are characteristic of still water ecosystems rich in invertebrates and amphibians with extensive macrophyte cover. For this reason, whether long-established or truly native, crucians can foster conservation support for the protection or restoration of small pond environments, attracting the involvement of local communities including roll-out across private land (Figure 3.11).

Crucian conservation then very much constitutes part of a cultural transition in understandings of a deeper, functional meaning of nature conservation. Nonetheless, many traditional conservationists still resist this changing reality, stuck in a conceptual rut of trying to 'put Humpty Dumpty together again' in a fundamentally and often irreversibly changed landscape. The changing face of conservation of nature and of the wider benefits of healthy and diverse ecosystems may nonetheless mean that crucians have important totemic roles to play in stemming the tide of habitat loss, ecosystem recovery, and in the wider journey of sustainable development.

Despite this, legacy nature conservation legislation does not identify crucians as a priority, unlike other freshwater fish species such as the Atlantic salmon, European eel and spined loach. The crucian is classified on the IUCN Red List of Threatened Species as a species of 'Least Concern' across its global range. However, this assessment was last updated in 2008 and was then accompanied by the caveat of a 'Declining' current population trend because of distribution contractions in its native range.[13] Across its European range, crucians have continued to decline due to competition from other species and habitat loss.

FIGURE 3.11   The appeal of crucians can create positive support for conservation of the small water bodies in which they thrive, along with other endangered small pond species. (Image © Mark Everard.)

The Bern Convention ('The Bern Convention on the Conservation of European Wildlife and Natural Habitats 1979'), the European Union (EU) Habitats Directive (Council Directive 92/43/EEC on the Conservation of Natural Habitats and of Wild Fauna and Flora) and United Kingdom nature conservation legislation (e.g. the Wildlife and Countryside Act 1981, as subsequently amended) do not list crucians for any special nature conservation attention (Figure 3.12). However, the Bern Convention and the EU Habitats Directive both impose bans on a number of listed destructive fishing methods.

Threats from Other *Carassius* Species

Aside from loss or pollution of suitable habitat, and competition or predation from other types of fishes, closely related invasive *Carassius* species have become a serious threat to crucian carp populations in Central-Eastern Europe and South-Eastern England through competition for space and food resources, and also hybridisation. The crucian's close relationship to goldfish (*Carassius auratus*), widely naturalised in Britain and across much of the world, poses threats related not merely to competition but also due to hybridisation.

Until (it was believed) 2021, gibel (*C. gibelio*) had not been perceived as a threat in Britain, though they were regarded as one of the most pernicious concerns for crucian populations across continental Europe. The continuing extirpation of crucians

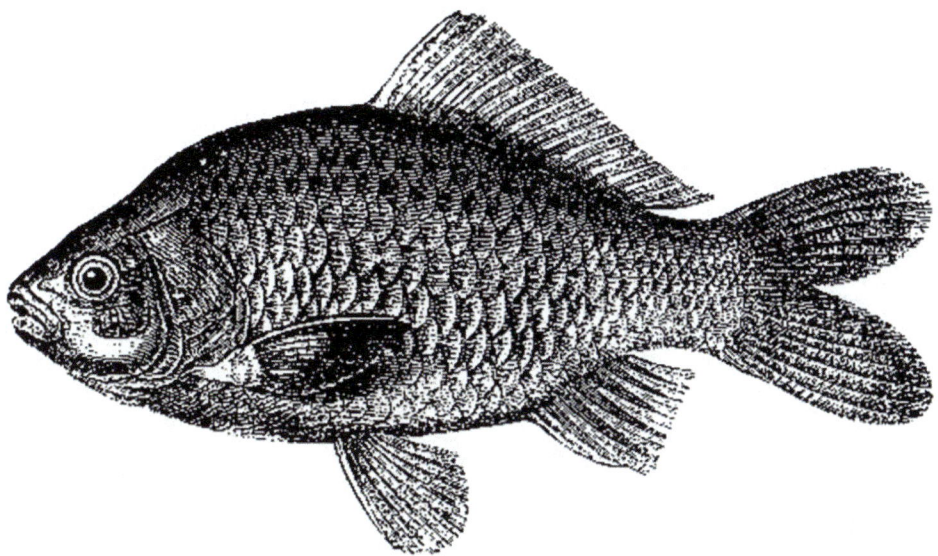

FIGURE 3.12   Old woodcut of a crucian, though with a questionably straight outer edge to the dorsal fin casting some doubt about genetic purity.

from many water bodies, especially in the Danube drainage and central Europe, are believed to be substantially due to competition from introduced gibel in non-optimal habitats.[14] Gibel have been found to outperform crucians in terms of growth rate and effectiveness of resource use in both field and laboratory experiments.[15] The island setting of Britain had formerly been a bulkhead for crucians – an 'ark' population – against competition from ever-spreading gibel that pose such a serious threat to crucians across Europe. However, though not previously a perceived problem in Britain, evidence of the arrival and local establishment of gibel in Britain in 2021 – and the likelihood that these fish have been present here undetected for far longer – changes that cosy assumption.

Now, we can add to threats from *Carassius* species the unpredictable consequences of establishment of nigorobuna (*Carassius grandoculis*) in British waters. We will have to see what problems gibel and nigorobuna invasion bring with them, whilst recognising the importance of continuing biosecurity to block any further unintended introductions or spread.

Crucian Conservation in Britain

Nature conservation interest in the crucian in Britain is mixed, much of it related to debate about the native status of the fish. Crucian remains date back to Roman Britain. As noted previously, a crucian bone was found in a Roman midden (a refuse heap generally associated with kitchens). This is not of itself proof that this fish was native, as these few remains could have been brought by travellers from continental Europe.

More definitive analyses of DNA from crucians found in England suggests a medieval introduction or reintroduction, roughly in the fifteenth century, though this date is an estimate. It is by no means certain whether this relates to a primary introduction, or whether this new stock was a reintroduction replacing dwindling pre-existing populations.[16] Perhaps the absence of mention of the crucian in Izaak Walton's *The Compleat Angler*, written in 1653, is a clue? And, as we have seen, Peter Rolfe suggests that 300 years ago is more likely as this is the time when goldfish were first brought to Britain. There is now general consensus in the scientific community that crucians are not strictly native to Britain as defined by being present here before the Doggerland bridge was breached at the end of the last ice age.

Whether truly native or not, crucians are nonetheless long-established in England. Crucian populations have also more recently been in sharp decline here. This, in itself, is not a compelling reason for conservation of the crucian as many other species have been introduced to Britain, including for example highly invasive species such as common carp, wels catfish (*Silurus glanis*), topmouth gudgeon (*Pseudorasbora parva*) and goldfish. However, the fact that the crucian is characteristic of small pond habitats (Figure 2.13) – a threatened resource in British landscapes – competing only poorly with other fish species but happily cohabiting

FIGURE 3.13 Crucians are characteristic and charismatic denizens of small, well-vegetated pond habitats that they share with a range of other vulnerable wildlife. (Image © Mark Everard.)

with other characteristic fauna and flora of these now-threatened habitats does, as we have seen, provide cultural focus for protection or restoration of ponds and other smaller wetlands.

A 2020 study[17] assessing the success of a decade of crucian introduction/re-introduction efforts in Norfolk recorded not only a 72% decline in crucian distribution between the 1950s–1980s and the 2010s, but also that pond rehabilitation and restocking since that time has achieved a substantial recovery. Other crucian conservation initiatives are taking place throughout England, including a breeding population of crucians in two of my own garden ponds enabling seeding of yearlings into suitable local ponds including those where crucians were formerly present. This recent localised renaissance in crucian populations is a recovery from a low base after decades of habitat loss – there has been an estimated 50% loss of farm ponds over the past 50 years,[18] exacerbated by agricultural, urban and other encroachment and pollution – with the pervasive stocking of common carp in remaining water bodies exerting further pressure.

Conflicting views about crucians are taken by England's environmental regulators. The Environment Agency fish hatchery at Calverton, near Nottingham, breeds DNA-tested crucians for stocking, and is a partner in the National Crucian Conservation Project (NCCP, described in detail later in this chapter). Conversely, Natural England, the national nature conservation regulator, regards the crucian as a non-native species and opposes its stocking in designated wildlife sites regardless of whether it was formerly present or in the light of compelling evidence that the species is entirely synergistic with other characteristic species found in these vulnerable habitats.

As noted previously, crucians are not scheduled under primary UK and European nature conservation legislation, other than bans on a number of listed destructive fishing methods under the Bern Convention and the EU Habitats Directive. However, in 2010, crucians were established as a Biodiversity Action Plan (BAP) species in Norfolk, their first formal conservation designation in the United Kingdom.[19] Nonetheless, crucians have received concerted voluntary conservation action in Britain over recent years. Significant amongst these are the NCCP and, on a more localised basis, the Norfolk Crucian Project. However, in the light of the roles of this fish in the profoundly changed landscape of modern Britain, there are also wider compelling reasons why crucian conservation is important for the habitats they depend upon.

The National Crucian Conservation Project

The genesis of the NCCP was the passion of artist and author Chris Turnbull, also in collaboration with Dr (now Professor) Carl Sayer from University College London (Figure 3.14). Both are 'Norfolk boys' at heart, Carl by birth and Chris since 1980, concerned that the rich pondscape of the county and its associated wildlife, including the crucian, was being lost at a frightening pace. Chris in particular drove wider interest in identifying where crucians populations remained, study of their genetic provenance

FIGURE 3.14   Logo of the National Crucian Conservation Project (NCCP). (With permission.)

and activities to restore their diminishing habitats. The Angling Trust came on board promoting the cause and, at its Coarse Fish Conference in Reading in May 2014, the Environment Agency joined forces to promote the NCCP. I am pleased to say that I was part of that meeting.

The NCCP is a body with no money and no employees. However, its strength has been in bringing together a wide range of supporters and knowledgeable representatives across a spectrum of societal sectors – angling, academic, voluntary organisations, members of the general public – sharing a common interest in furthering the status of the crucian. The twin objectives of the NCCP are to: promote the conservation of the species and its habitat, and encourage the development of well-managed crucian fisheries. This twin-track approach seeks to reverse the decline in crucian habitat and promote accredited 'true crucian' fisheries free from crucian-like fishes of uncertain genetic provenance, including hybrids.

New crucian waters have since been established in many parts of the country and a number of pre-existing crucian fisheries have been improved and rejuvenated. A list of recognised waters can be found on the NCCP website, hosted by the Angling Trust.[20] The NCCP has also developed a 'crucian catalogue' listing details of crucian-containing waters spread across the British Isles graded into four categories:

- A. *'Managed as Premium Crucian Fishery'*: In these fisheries, the crucian stock is recognised as being true crucians, generally DNA certified with no F1s or goldfish present and common carp either absent or interbreeding managed.
- B. *'Mixed Fishery with Stock of Crucians'*: These waters contain crucian stocks supplied by the Environment Agency or from an approved

supplier, or that have been photo-checked to assess their authenticity. Again, no F1s or goldfish are present and interbreeding with common carp is managed.

- *C. 'Mixed Stock Fishery'*: These fisheries contain genuine crucians, but may also contain F1s, common carp and/or goldfish or hybrids. Any purported crucians caught from these waters can pose identification problems, with crucian captures likely to be occasional.
- *D. 'Ungraded'*: This last category list waters claim to contain crucians, though with recent captures either rare or unverified, and with an unknown stocking history. There is a likelihood that these waters may no longer contain crucians or that fish claimed to be crucians are potentially F1s, goldfish or hybrids. Further investigation is required before the venue can be allotted a higher grading.

Peter Rolfe remains a font of knowledge and has helped guide the efforts of the NCCP. The Environment Agency's National Coarse Fish Rearing Unit at Calverton, near Nottingham, also raises DNA-verified crucians for stocking into dedicated waters.

The Norfolk Crucian Project

The Norfolk Crucian Project was established in 2009 in response to the decline of the crucian, and particularly since a 2008 Environment Agency announcement that the species was *"…thought to be virtually extinct in Norfolk"*. This declining trend appeared not only among the crucian's stronghold in eastern England, but also widely in farmland ponds across Britain mirroring trends across large tracts of Europe. In collaboration with Cefas (the UK government's Centre for Environment, Fisheries and Aquaculture Science), the Pond Restoration Research Group at University College London has undertaken fyke net surveys of many of Norfolk's ponds, including some known to have contained the species in the 1970s–1980s.

Professor Carl Sayer, the prime instigator of the Norfolk Crucian Project, wrote in the 2021 book *Willow Pitch VI: Crucian Renaissance*

> *Right from the beginning, the aim of the Norfolk Crucian Project has remained singular and clear: that when we are old, there will be crucians in the ponds of Norfolk. That aim has never changed.*

Furthermore, a *Ghost Ponds Project* is focused on re-excavation of lost ponds – 'ghost ponds' – that have been filled in for agricultural land reclamation especially since the 1950s. Buried within the historic sediments of these former ponds lies a seed bank of past native aquatic plants, which can regenerate in favourable conditions to recolonise these revived waters.

Wider Reasons for Crucian Conservation in Contemporary Britain

The twin crises of biodiversity loss and climate change are global phenomena; 8.3 billion people are now living in an environment that is not only radically changed but will inevitably be subject to further change. This problem is far from remote; the United Kingdom is now the most nature-depleted country in the world as a result of a long history of unsympathetic agricultural and industrial development as well as growing population density.[21]

This raises challenging questions about the objectives of nature conservation. The paradigm under which nature conservation has largely operated since the middle of the twentieth century has been one of preserving habitats and species native to the regions in which they have occurred. In an overpopulated world in which humans have radically altered not only the structure, chemistry and hydrology of landscapes and waterscapes but also the climate in which they reside, a rather simplistic aspiration to 'put Humpty-Dumpty back together again' may now be entirely infeasible. Moreover, shifting phenology – the timings and synchronisation of natural occurrences – may result in the breakage of important ecological links, such as timely availability of fine food items for emerging fish larvae or nesting birds, and adequately high water levels essential for the breeding, feeding, refuge and other needs of multiple species (Figure 3.15).

FIGURE 3.15   Charismatic appeal, heritage attributes and benign relationship with other small pond wildlife means that crucians can play important roles promoting conservation of these threatened habitats. (Image © Mark Everard.)

There is increasing recognition of the difficulties inherent in re-creating past eco-systems given the scale of contemporary climatic shifts and landscape change. As we have seen, these factors are driving a new focus on ecosystem complexity and func-tioning, and on working with natural processes rather than simply seeking to replace expected species as a foundation for ecosystem restoration. The hard reality is that replacing former species may be impossible or impractical if the environmental con-ditions required for them to complete their life cycles, as well as those of the other species with which they interact as elements of complex ecosystems, are no longer amenable. We have to turn our attentions instead to the roles that species play in the functioning of ecosystems.

Non-invasive species such as crucians can play key roles, whether strictly native or not. They have charismatic and heritage attributes forming a focus for re-evaluation of the importance of small pond habitats. They are also well adapted to cohabit with and are benign to other characteristic wildlife of these vulnerable and significantly declining pond habitats. Even without this broader philosophical thinking, maintain-ing crucian strongholds is important as the species continues to decline nationally, pond conservation also serving to increase overall landscape-scale biodiversity. Add to this that deepening evidence of climate change impacts emphasises the need for greater investment in the resilience of freshwater bodies and their significant roles in whole ecosystem functioning, including rehabilitation or restoration of lost or degrad-ing habitats.

Further importance concerning the maintenance of crucian strongholds in Britain is that the species continues to decline precipitously across its formerly extensive European range, due both to land use changes and the relentless spread of gibel. Britain is now effectively a global stronghold for the crucian, protected by the surrounding sea though itself also recently invaded by gibel.

PET CRUCIANS

The small 12-inch (30-centimetre)-long aquarium in my study at home houses three or four small crucians from the spawning from two years ago, overwintering indoors some of the stock from my breeding pond though most of the progeny from that year's spawning grow on outside in a separate pond. The crucians both in my study tank and another tank in the living room grow on strongly and, perhaps surpris-ingly, rapidly become tame in close proximity to people. As the photograph in Figure 3.16 indicates, some of my previous tank occupants were taking food from my fin-gers within 36 hours of me bringing them in from the pond in which they had been spawned and spent their entire lives up to that point. Aside from enjoying sharing our living space with these fascinating fish and observing their behaviours at close quarters, this refuge proved useful when the need arose to replace some of the brood stock in my breeding pond after a raiding otter visited the garden during the winter, clearing all of its occupants.

FIGURE 3.16   These wild-hatched crucians surprisingly were taking flaked food from my fingers within 36 hours of being brought indoors into my study fish tank. (Image © Mark Everard.)

Add their hardiness to their adaptability and tameness, and that they are also unfussy eaters feeding freely on dried aquarium fish flake food, and we have a species that has all the attributes of a good aquarium fish. This is another aspect of their close kinship with goldfish. As Alwyne Wheeler wrote of the crucian in his 1969 book *The Fishes of the British Isles and North West Europe*

> *It makes an attractive aquarium fish, and quickly becomes surprisingly tame.*

Although I have a breeding population of crucians in my garden ponds, the juveniles of which I give away for stocking suitable adjacent ponds, crucians are less ideal as an ornamental garden pond fish. Though they are remarkably hardy and they also do

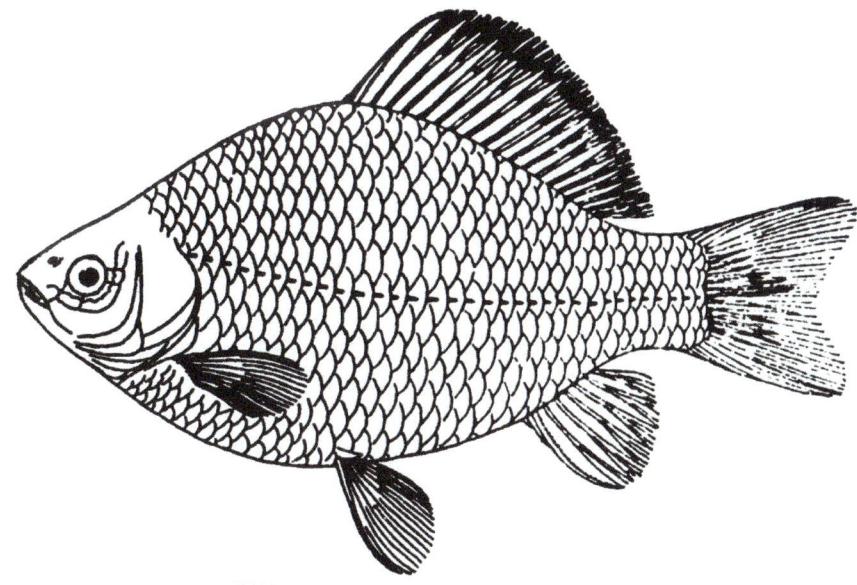

FIGURE 3.17   Drawing of crucian by Tate Regan, 1911.

not outgrow the pond, you rarely see the adults other than when picked out by a power-ful spotting torch in the dark hours. As Lawrence Wells wrote in his 1941 book *The Observer's Book of Freshwater Fishes of the British Isles,*

> *As a fish for the garden pond it has little value preferring to lurk about the mud and plant roots on the bottom.*

The exception to this is when they get frisky during summertime, spawning multiple times, when you might see one jump or flip over rapidly as they rush to the surface and as quickly dive for the bed. The fry and juveniles though may be seen moving around, including browsing for small food items in the surface film on warmer days.

That said, crucians are hardy and prolific, and a fish to satisfy the enthusiast. It goes without saying that they should not be mixed with goldfish, common carp or gibel – or now the newly arrived nigorobuna – to avoid the generation and accidental spread of hybrids (Figure 3.17).

THE ECONOMICS OF CRUCIANS

Where present in mixed fish stocks, crucians can make important contributions to harvestable natural resources of freshwater ecosystems. For example, they are cited in the 2013 Annual Report of the Danube Delta Biosphere Reserve (see Bibliography) under the heading 'Natural resources utilization, traditional activities', which states:

> The main renewable natural resources used in 2013 were the fish and the reed. The fish capture recorded a value of around 1,312 tones and was dominated by fresh water species: Abramis brama, Carassius carassius, Rutilus rutilus, Aspius aspius, Carassius gibelio, Cyprinus carpio, Esox lucius, Perca fluviatilis, Silurus glanis, Stizostedion lucioperca and Blicca bjoerkna. In the 2012–2013 harvesting season, the quantity of reed harvested was of 3,019 tones.

Bent J Muus and Preben Dahlstrom commented in their 1967 book *Collins Guide to the Freshwater Fishes of Britain and Europe* that

> In open waters the crucian carp is mainly fished in traps, and there is some interest in this species among anglers.

In his 1995 *Multilingual Illustrated Guide to the World's Commercial Coldwater Fish*, C Frimodt documented that crucians are marketed fresh and frozen, and are eaten fried, broiled and baked across their geographical range.

Crucians can also be, or at least have been, important in aquaculture. In his 1969 book *The Fishes of the British Isles and North West Europe*, Alwyne Wheeler notes that

> The Crucian carp is of considerable economic importance in eastern Europe, and is raised in fish farms, and released in poor-quality artificial waters to be cropped later. It is also a valuable angling fish, not least for its ability to thrive in places where few other species could survive....
>
> It has been widely introduced both in the British Isles and Europe.

CRUCIAN SOCIETIES

Many freshwater fish species have associated societies. If you are interested in salmon and trout, pike or barbel, carp, tench or perch, roach and pretty much any other fish species, including such tiddlers such as gudgeon and ruffe, you are well served by dedicated societies.

Crucians too have their adherent enthusiasts, the more so in recent years with the rising profile of the species, including conservation activities and the increase in accredited fisheries. One such body is the virtual *Association of Crucian Anglers*, operating mainly via its Facebook (social media) page https://www.facebook.com/groups/1463582757197274.

The NCCP is an association of sorts, lacking money and staff so essentially comprising a cross-sectoral network of interested and engaged individuals and organisations.

Perhaps more crucian-related organisations will become established as the profile of and interest in this charismatic little fish continues to grow.

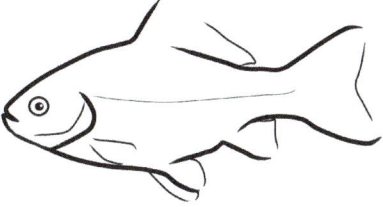

NOTES

1. Sidechef.com. *Chinese Crucian Carp Soup.* https://www.sidechef.com/recipes/1704/ Chinese_Crucian_Carp_Soup.
2. SimpleChineseFood.com. *Beer Crucian Carp.* https://simplechinesefood.com/recipe/ beer-crucian-carp.
3. Strybel, R. and Strybel, M. (2005). *Polish Heritage Cookery.* Hippocrene Books, New York.
4. Molokhovets', E. (1998). *Classic Russian Cooking: Elena Molokhovets' a Gift to Young Housewives.* Indiana University Press, Bloomington.
5. Volokh, A. and Manus, M. (1983). *The Art of Russian Cuisine.* Macmillan, London.
6. FAO. (2009). *Carassius carassius.* In Cultured aquatic species fact sheets. Food and Agriculture Organization of the United Nations (FAO), Rome. [Online.] https://www.fao. org/fishery/docs/DOCUMENT/aquaculture/CulturedSpecies/file/en/en_cruciancarp.htm, accessed 21 February 2024.
7. Lawton, J. (2010). *Making Space for Nature: A review of England's Wildlife Sites and Ecological Network.* Department for Environment, Food and Rural Affairs (Defra), London. [Online.] https://webarchive.nationalarchives.gov.uk/20130402170324/http://archive. defra.gov.uk/environment/biodiversity/documents/201009space-for-nature.pdf, accessed 8 February 2024.
8. WWF. (2021). *One-third of freshwater fish face extinction and other freshwater fish facts.* WWF. [Online.] https://www.worldwildlife.org/stories/one-third-of-freshwater-fish-face-extinction-and-other-freshwater-fish-facts, accessed 9 February 2024.
9. Deinet, S., Scott-Gatty, K., Rotton, H., Twardek, W. M., Marconi, V., McRae, L., Baumgartner, L. J., Brink, K., Claussen, J. E., Cooke, S. J., Darwall, W., Eriksson, B. K., Garcia de Leaniz, C., Hogan, Z., Royte, J., Silva, L. G. M., Thieme, M. L., Tickner, D., Waldman, J., Wanningen, H., Weyl, O. L. F. and Berkhuysen, A. (2020). *The Living Planet Index (LPI) for migratory freshwater fish – Technical Report.* World Fish Migration Foundation, The Netherlands.
10. Biggs, J., von Furnetti, S. and Kelly-Qunn, M, (2017). The importance of small waterbodies for biodiversity and ecosystem services: implications for policymakers. *Hydrobiologia,* 793, pp. 3–39. DOI: https://doi.org/10.1007/s10750-016-3007-0.
11. Williams, P., Biggs, J., Crowe, A., Murphy, J., Nicolet, P., Weatherby, A. and Dunbar, M. (2010). *Countryside Survey: Ponds Report from 2007.* CS Technical Report No.7/07. https://nora.nerc.ac.uk/id/eprint/9622/1/N009622CR.pdf.
12. State of Nature Partnership. (2023). *State of Nature 2023.* State of Nature Partnership. https://stateofnature.org.uk/wp-content/uploads/2023/09/TP25999-State-of-Nature-main-report_2023_FULL-DOC-v12.pdf.

13. Freyhof, J. and Kottelat, M. (2008). *Carassius carassius*. The IUCN Red List of Threatened Species 2008: e.T3849A10117321. https://dx.doi.org/10.2305/IUCN.UK.2008.RLTS.T3849A 10117321.en.

14. Kottelat, M. and J. Freyhof, 2007. *Handbook of European freshwater fishes*. Publications Kottelat, Cornol and Freyhof, Berlin. 646 pp.

15. Tapkir, S.D., Boukal, D., Kalous, L., Bartoň, D., Souza, A.T., Kolar, V., Soukalová, K., Duchet, C., Gottwald, M. and Šmejkalet, M. (2022). Invasive gibel carp (*Carassius gibelio*) outperforms threatened native crucian carp (*Carassius carassius*) in growth rate and effectiveness of resource use: field and experimental evidence. *Aquatic Conservation Marine and Freshwater Ecosystems*, 32(12), pp. 1901–1912. DOI: https://doi.org/10.1002/ aqc.3894.

16. Jeffries, D.L., Copp, G.H., Maes, G.E., Lawson Handley, L., Sayer, C.D. and Hänfling, B. (2017). Genetic evidence challenges the native status of a threatened freshwater fish (*Carassius carassius*) in England. *Ecology and Evolution*, 7, pp. 2871–2882. DOI: https:// doi.org/10.1002/ece3.2831.

17. Sayer, C.D., Emson, D., Patmore, I.R., Greaves, H.M., West, W.P., Payne, J., Davies, G.D., Tarkan, A.S., Wiseman, G., Cooper, B., Grapes, T., Cooper, G. and Copp, G.H. (2020). Recovery of the crucian carp *Carassius carassius* (L.): Approach and early results of an English conservation project. *Aquatic Conservation: Marine and Freshwater Ecosystems*, 30(12), pp. 2240–2253. DOI: https://doi.org/10.1002/aqc.3422.

18. WWT. (2022). *Restoring lost farmland ponds*. Wildfowl and Wetlands Trust (WWT). [Online.] https://www.wwt.org.uk/our-work/projects/restoring-lost-farmland-ponds, accessed 23 September 2022.

19. Copp, G.H. and Sayer, C.D. (2010). Norfolk Biodiversity Action Plan – Local Species Action Plan for Crucian Carp (*Carassius carassius*). Norfolk Biodiversity Partnership. Reference: LS/3. Lowestoft, Suffolk: Centre for Environment, Fisheries & Aquaculture Science. http:// www.cefas.defra.gov.uk/publications/files/Copp-Sayer2010_CrucianCarpBAP_12-02-10. pdf.

20. NCCP. *National Crucian Conservation Project*. Angling Trust. https://anglingtrust.net/ national-crucian-conservation-project/.

21. State of Nature Partnership. (2023). *State of Nature 2023*. State of Nature Partnership. https://stateofnature.org.uk/wp-content/uploads/2023/09/TP25999-State-of-Nature-main-report_2023_FULL-DOC-v12.pdf.

Bibliography

The following works are referenced in this book, with my thanks to the authors concerned where quoted.

American Physiological Society. (2006). Remarkable physiology allows crucian carp to survive months without oxygen. *ScienceDaily*, 26 August 2006. [Online.] www.sciencedaily.com/releases/2006/08/060825103548.htm, accessed 26 December 2022.

Angling Times. (2020). British record crucian carp – the full story. *Angling Times*, 06 August 2020. [Online.] https://www.anglingtimes.co.uk/news/stories/british-record-crucian-carp-the-full-story/, accessed 27 December 2022.

Angling Times. (2021). British crucian record smashed! *Angling Times*, 4 May 2021. https://www.anglingtimes.co.uk/news/stories/british-crucian-record-smashed/, accessed 3 January 2024.

Bailey, John. (1992). *Fisherman's Valley: Seasonal Tips for Coarse Anglers*. David & Charles, Newton Abbot.

Bloch, Marcus-Eliezer. (1782 to 1795). *Allgemeine Naturgeschichte der Fische* (in 12 volumes).

Bloch, Marcus-Elieser. (1782). *Oeconomische Naturgeschichte der Fische Deutschlands. Erster Theil*. Hesse, Berlin.

Couch, Jonathan. (1877). *A History of the Fishes of the British Islands*. George Bell & Sons, London.

(The) Crucian Chronicle. (2017). The 1st Annual Newsletter and Journal of the Association of Crucian Anglers. The Association of Crucian Anglers.

Danube Delta Biosphere Reserve Authority. (2013). *Danube Delta Biosphere Reserve Authority: Annual Report for 2013*. Ministry of Environment and Climate Change, Romania.

Everard, Mark. (2013). *Redfin Diaries: A Life in the Year of a Roach Enthusiast*. Coch-y-Bonddu Books.

Everard, Mark. (2013). *Britain's Freshwater Fishes*. Princeton University Press/WildGUIDES.

Everard, Mark. (2020). *The Complex Lives of British Freshwater Fishes*. CRC/Taylor and Francis.

Fort, Tom. (2020). *Casting Shadows: Fish and Fishing in Britain*. William Collins, London.

Frimodt, C. (1995). *Multilingual Illustrated Guide to the world's Commercial Coldwater Fish*. Fishing News Books, Osney Mead, Oxford, England. 215pp.

Giles, Nick. (1994). *Freshwater Fish of the British Isles: A Guide for Anglers and Naturalists*. Swam Hill Press, Shrewsbury.

Günther, Albert. (1859-1870). *Catalogue of the Fishes in the Collection of the British Museum*. British Museum, London.

Gupta, N. and Everard, M. (2019). Non-native fishes in the Indian Himalaya: an emerging concern for freshwater scientists. *International Journal of River Basin Management*, 17 (2), pp. 271–275. https://doi.org/10.1080/15715124.2017.1411929.

Houghton, MA, FLS, The Reverend W. (1879). *British Fresh-Water Fishes*. William Mackenzie, London. (Note: This book has been reprinted over the decades by numerous publishers, for example by The Peerage Press, London, in 1981.)

Jeffries, D.L., Copp, G.H., Maes, G.E., Lawson Handley, L., Sayer, C.D. and Hänfling, B. (2017). Genetic evidence challenges the native status of a threatened freshwater fish (Carassius Carassius) in England. *Ecology and Evolution*, 7, pp. 2871–2882. DOI: https://doi.org/10.1002/ece3.2831.

Karvonen, A., Bagge, A.M. and Valtonen, E.T. (2005). Parasite assemblages of crucian carp (*Carassius Carassius*) – Is depauperate composition explained by lack of parasite exchange, extreme environmental conditions or host unsuitability? *Parasitology*, 131(2), pp. 273–278. DOI: https://doi.org/10.1017/S0031182005007572.

Keith, P. and Allardi, J. (2001). *Atlas des poissons d'eau douce de France. Muséum national d'Histoire naturelle, Paris. Patrimoines naturels*, 47, pp. 1–387. DOI:

Kottelat, M. and Freyhof, J. (2007). *Handbook of European Freshwater Fishes*. Publications Kottelat, Cornol and Freyhof, Berlin. 646 pp.

Lacépède, Bernard Germain. (1798–1803). *Illustrations de Histoire naturelle des poissons. Marie-Anne Rousselet (engraver)*. Chez Passan, Paris, France.

Maitland, Peter and Campbell, R. Niall. (1992). *Freshwater Fishes*. Collins, London.

Marshall-Hardy, Eric. (1943). *Coarse Fish*. Herbert Jenkins, London.

Muus, Bent J. and Dahlstrom, Preben. (1967). *Collins Guide to the Freshwater Fishes of Britain and Europe*. Collins, London.

Newdick, J. (1983). *The Complete Freshwater Fishes of the British Isles*. A&C Black (Publishers) Ltd., London.

Pinder, A.C. (2001). *Keys to Larval and Juvenile Stages of Coarse Fishes from Fresh Waters in the British Isles*. Freshwater Biological Association Scientific Publications Volume 60. Freshwater Biological Association, Windermere.

Rolfe, Peter. (2010). *Crock of Gold: Seeking the Crucian Carp*. MPress, Romford.

Rolfe, Peter. (2023). *Old Angler Rambling*. Kindle Direct Publishing.

Rolfe, Peter. (2024). *Crock of Gold: The Crucian Revealed*. Peter Rolfe, Shaftesbury.

Sterba, Günther. (1959). *Freshwater Fishes of the World*. Vista Books Ltd, London.

Stewart, Charles. (1817). *Elements of the Natural History of the Animal Kingdom*. Bell and Bradfute; and Longman, Hurst, Rees, Orme, & Brown.

Turnbull, Chris (editor). (2021). *Willow Pitch VI: Crucian Renaissance*. Little Egret Press/ Water's Edge Publications.

Walton, Izaak and Cotton, Charles. (1653). *The Compleat Angler*. Maurice Clark, London. (Available these days in many editions and from various publishers).

Wheeler, Alwyne. (1969). *The Fishes of the British Isles and North West Europe*. Michigan State University Press.

Wells. A. Lawrence. (1941). *The Observer's Book of Freshwater Fishes of the British Isles*. Frederick Warne and Co. Ltd.

WWT. (2022). *Restoring lost farmland ponds*. Wildfowl and Wetlands Trust (WWT). [Online.] https://www.wwt.org.uk/our-work/projects/restoring-lost-farmland-ponds, accessed 23 December 2022.

Yarrell, William. (1859). *A History of British Fishes in Two Volumes*. John Van Voorst, London.

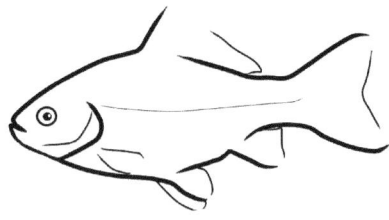

Index

Note: Page numbers in *italics* and **bold** refer to figures and tables, respectively.

A

Abramis brama see common bream
acclimatisation society 15
Acheilognathinae (bitterling-like cyprinids) 25
Allen, Jay 62
Allgemeine Naturgeschichte der Fische (Bloch) 77
amorphous matter 18
amphibians 81
anglers 17, 27, 41–46, 48, 50, 52, 54, 58, 62, 74, 77
angling 14, 27–28, 41–46, 49, 65, 74–75; authors 18; branches of 46; Britain 27, 57; classification of 42; records 60; recreational 14, 28; specimen scene 41
Angling Times 12, 60–61
Angling Trust 33, 60, 86
anoxia 8–9, 11
aquaculture 25, 64, 73–74, 91
aquatic environments 79
aquatic insects 46
aquatic invertebrates 16
aquatic plants 8, 16, 87
asafoetida 50
Aspius aspius 91
Association of Crucian Anglers 92
Atlantic salmon 15, 57, 76, 81, 92
Australia 15, 16, 65

B

back-crossing 36–37
Bailey, John 43, 52, 57
baits 45–52; boilies/synthetic 51; bread 48–50, *49*; insect larvae 45–46 (*see also* maggots); invertebrate 46–48; loose feed 52; meaty 47–48; plant/organic 50–51
BAP *see* Biodiversity Action Plan (BAP)
barbel (*Barbus barbus*) 11
Barbinae 25
Barnes, Julian 60, *61*
Beer Crucian Carp, recipe 72
Bern Convention 82, 85
Berry, Wm. 62
biodiversity 10, 79–80, 88
Biodiversity Action Plan (BAP) 85

Blavins, Joshua 61
Blicca bjoerkna 91
Bloch, Marcus-Eliezer 12, 31, 77, *77*
Bloodworms *17*, 46
boilies 45, 50, 51
bolt rig 54
borshch s karasei 73
Bowler, Martin 62
bread baits 48–50, *49*
bream carp 2, 69
British crucians 11–14
British Fresh-Water Fishes (Houghton) 2, 8, 12, 26, 31, 72, 77
British Record (Rod Caught) Fish Committee (BR(RC)FC) 60–62
Brooke Bond company 77, *78*
brown goldfish *see* goldfish (*Carassius auratus*)
brown trout (*Salmo trutta*) 15, 16, 60
BR(RC)FC *see* British Record (Rod Caught) Fish Committee (BR(RC)FC)
bryophytes 16
bullheads 15
burbot (*Lota lota*) 11

C

Campbell, R. Niall 13
Carassius langsdorfii 26
Carassius praecipuus 26
Cardozo, Peter 61
carp (*Cyprinus carpio*) 1, 7, 25–36, 27, *71*; Beer Crucian 72; common 27–29, *28*, **34–35**, 36, 64, 91; German 2, 26, 69; Hamburgh 2, 26, 69; minnow 15, 25–26, 57; Prussian 12, 17, 31, 76, *76*; societies 13–14, 27–29, *28*, **34–35**, 36, 44, 47, 52, 64, 84, 91
casters 45–46, 50
Casting Shadows: Fish and Fishing in Britain (Fort) 27
Catalogue of the Fishes in the Collection of the British Museum (Günther) 12
climate change 80, 88
Coarse Fish (Marshall-Hardy) 12, 18, 21–22, 28, 62

Collins Guide to the Freshwater Fishes of Britain and Europe (Muus and Dahlstrom) 8, 9, 11, 18–21, 24, 28, 73, 91
colouration 4–5, 49
common bream (*Abramis brama*) 5, 15, 19, 26, 32, 47, 52, 91
common carp (*Cyprinus carpio*) 27–29, *28*, **34–35**, 36, 64, 91; physical features of 28
The Compleat Angler (Walton) 74, 84
The Complete Freshwater Fishes of the British Isles (Newdick) 7, 13, 18–19
Couch, Jonathan 12, 31
Covid 60
creative arts 76–79
Crock of Gold: Seeking the Crucian Carp (Rolfe) 18, 24, 37, 63, 66, 75
Crock of Gold: The Crucian Revealed (Rolfe) 37, 66, 75
crucian bone, discovery of 13
crucian (*Carassius carassius*) *1*, 1–37, *23*, *41*, *73*, *75*, *81*, *83–84*, *90*; *see also* carp (*Cyprinus carpio*);
in aquaculture 73–74; as bait *47*, 64; biology 37; British 11–14; and creative arts 76–79; cuisine 72–73; diet of 17–19, 47; different meanings of *68*, 69–72; ecology 42; economics of 91–92; etymology 69; fight of 57–58; fishing (*see* fish/fishing);
habits/habitats 7–8, *84*; head/mouth of 5, *6*, *29*; known names of **70**, **71**; life cycle 19–22; in literature 74–76; location 43–44; low-oxygen conditions 8; match catches 64; more information 37; natural distribution of 10–11; nature conservation 79–89; painting of *76–78*, *92*; and people 68–92; pests/diseases 24–25; pet 89–91; photographing 58–60; physical features of 2–3, **34–35**; population 13–14, 24, 33, 42, 65–66, 82, 85, 90; predation/shape 22–24; as problems 64–65; records 60–63; roles 81, 88, *89*; societies 92; spread of 14–15; superpowers 8–10; taxonomy 25–37
crucian carp (*Cyprinus carassius*), defined 1
cuisine 72–73, *73*
Cyprinidae family 25–26; *see also* carp
Cyprinus Hamburger 2, 12

D

Dahlstrom, Preben 8–9, 11, 13, 18–21, 24, 28, 73, 91
Danioninae 25
Danube Delta Biosphere Reserve 91
Dendrobaena (red worms) 47

detritus, defined 18
'Die Forelle' (Schubert) 76
DNA analysis/testing 4, 36, 65
Doggerland, defined 11, 15, 84
Dungeons & Dragons (game) 72

E

East Han Dynasty 74
ecosystem 15–16, 18, 79–81, 88, 91
egrets 22–23
Elements of the Natural History of the Animal Kingdom (Stewart) 13
endozoochory 16–17
Enton Lake *see* Johnson's Lake
Environment Agency 30, 33, 77, *78*, 85–87
Esox Lucius 92, 23
etymology 69
EU Habitats Directive 82, 85

F

fauna 11, 13, 15, 85
feeder method *51*
FISHBASE lists 10, 26
Fisherman's Valley: Seasonal Tips for Coarse Anglers (Bailey) 43, 52, 57
The Fishes of the British Isles and North West Europe (Wheeler) 17, 30, 62, 90, 91
fish/fishing 41–67; angling 14, 18, 27–28, 41–46, 49, 60, 65, 74–75; baits 45–52, 64; culinary 33; ecology 42; fight 57–58; former times records 62–63; freshwater 12, 79; location 43–44; lure 57; match catches 64; modern times records 60–62; photographing 58–60; post-carp environment 65; presentation 52–56; problems 64–65; running crucian 65–67; summer 44–45
Food and Agriculture Organization (FAO) 64, 74
former times records 62–63
Fort, Tom 27
fortress conservation model 79
Frapwell, Steve 61
freshwater fishes, roles of 79
Freshwater Fishes (Maitland and Campbell) 13
Freshwater fishes of the world (Sterba) 11
Freshwater Fish of the British Isles: A Guide for Anglers and Naturalists (Giles) 7, 13–14, 18
Frimodt, C. 91

G

gardon 15, **70**; *see also* roach
geographical distribution 10–11

German carp 2, 26, 69
Ghost Ponds Project 87
gibel (*Carassius gibelio*) 24, 26–27, 29, 31–33, *32*, **34–35**, 64, 74, *81*, 91; *see also* Prussian carp
Giles, Nick 7, 13–14, 18
Giraldus Cambrensis 15
Gobioninae/Gobionidae 25, 26; *see also* gudgeon
Godalming Angling Association 60, 65
goldfish (*Carassius auratus*) 1, 12–14, 24, 26–27, 29, *30*, 30–31, **34–35**, 44, 57, 64, 69, 80, 82, 84, 86–87, 90–91
gudgeon 15, 25, 26, 57, 84, 92
Günther, Albert 12

H

habits/habitats 43, 80, 85–86, *89*; crucian 7–8; pond *1*, 25, 81, 84, *84*, 88; wetland 81, 85
Hamburgh carp 2, 26, 69
herons 22–23
Hinson, H.C. 62
The History and Topography of Ireland (Wales) 15
A History of British Fishes in Two Volumes (Yarrell) 12
Houghton, Reverend W 2, 8, 12, 26, 31, 72, 77
How to Catch Them (Jenkins) 74
hybrids/hybridisation 5, 7, 28, **34–35**, 36–37, 64, 75, 82; carp/crucian carp 28, 33; features of 29; goldfish/crucians 13

I

'ichthyologists' 12, 31
India 16
insect larvae 45–46
internal features crucian 7, **34–35**
Ireland 15–16
IUCN Red List 81

J

James, Michael 61
Japanese white crucian carp (*Carassius cuvieri*) 26
Jenkins, Herbert 74
Johnson's Lake 60–61, 65
juvenile development phases 21, *22*

K

Karusse 69, **70**
Keys to Larval and Juvenile Stages of Coarse Fishes from Fresh Waters in the British Isles (Pinder) 21
kingfishers 22–23

L

Lacépède, Bernard Germain 2, 12, 26, 31
larvae, characteristics of 21, 45
La Societé Zoologique d'Acclimatation 15
lateral line/scales 3–4
Leptobarbinae 25
Leuciscinae/Leuciscidae 25, 26
life cycle 19–22
lift method 53–54, *54*
Linnaeus, Carl 2, 12, 69
literature 74–76
Little Egret Press/Water's Edge Publications 76
loach 15, 81
Lobworms 46–47
loose feed 52
luncheon meat 45, 47–48
lure fishing 57; *see also* fish/fishing
Lydon, Alexander Francis *76*, 76–77

M

maggots *45*, 45–46, 50, 63
Maitland, Peter 13
Marshall-Hardy, Eric 12, 18, 22, 28, 62
Marsh Farm complex 60, 65
meaty baits 47–48
medieval clergyman 15
Miller, David 77, *78*
Milton Abbas Fishery 60–61
minnow/carps 15, 25–26, 57
mirror carp, defined 28
modern times records 60–62
mother ponds, defined 66
mudfish *see* crucian
Multilingual Illustrated Guide to the World's Commercial Coldwater Fish (Frimodt) 91
Muus, Bent J 8–9, 11, 13, 18–21, 24, 28, 73, 91

N

National Crucian Conservation Project (NCCP) 33, 43, 65–66, 85–87, *86*, 92
natural resources, utilization of 91
nature conservation 79–89; approaches to 79–80; in Britain 83–85, 88–89; NCCP 85–87; shifting approaches 80–83; thinking/ approaches to 79–80; threat 82–83
NCCP *see* National Crucian Conservation Project (NCCP)
Newdick, J. 7, 13, 18, 19
nigorobuna (*Carassius grandoculis*) 27, 29, 33, **34–35**, *36*, 64, 80, 83
Nilgiris (literally 'blue mountains') 16
Norfolk Crucian Project 85, 87
Norfolk Trading Standards 60

O

The Observer's Book of Freshwater Fishes of the British Isles (Wells) 14, 91
Old Angler Rambling (Rolfe) 1, 14, 75
old warrior, of crucian *63*

P

Paedocypridinae 25
Palmer, E. 62
pathogenic/non-pathogenic microbes 16
pellets 44, 45, 48, *48*, 49
perch (*Perca fluviatilus*) 11, 15, 17, 23–24, 46, 52, 57, 64–66, 77, 91, 92
pests/diseases 24–25
pet crucians 89–91
phenotypic plasticity 3, 10, 23
photosynthesis 8
physical features: colouration 4–5; fin/spines/rays 5–6; head/mouth 5; internal 7; lateral line/scales 3–4; species/hybrids 7
pike (*Esox lucius*) 15, 22, 23–24, 57, 64, 66, 91, 92
Pinder, A.C. 21
plant/organic baits 50–51
Polish Air Force 72
polyandry, defined 19
pond habitats *1*, 25, 81, 84, *84*, 88
Pond Restoration Research Group 87
population 19, 21, 48, 64, 80–81; crucian 13–14, 24, 33, 42, 65–66, 82, 85, 90; density 88; gibel 32–33; goldfish *30*; human 79; low-density 46
post-carp environment 65
predation/shape 22–24
presentation 42, 47, 49, 51, 52–56
Prussian carp (*Carassius gibelio*) 12, 17, 31, 76, *76*
PZL 23-A 'Karaś' 72

R

ray-finned fishes (Actinopterygii) 25
rays in fins 5–6
Redfin Dairies (Everard) 56
Regan, Tate 92
rehabilitation/restoration 85, 88
reoxygenation 8
roach (*Rutilus rutilus*) 15, 19, 26, 32, 46, 50, 52, 56, 57, 65, 66, 91, 92
Rolfe, Peter 1, 14, 18, 24, 31–32, 37, 63, 66, 75, 84, 87
rudd 15, 33, 46, 52, 56, 57, 65, 66
ruffe 57, 92
rule of thumb 44
running crucian fishery 65–67

S

Sayer, Carl 85, 87
Schubert, Franz 76
Second World War 62
Smith, Philip 62
Smithson, Craig 60–61
societies 92; carp 13–14, 27–29, *28*, **34–35**, 36, 44, 47, 52, 64, 84, 91; gudgeon/ruffe 15, 57, 84, 92; pike/barbel 15, 22, 23–24, 57, 64, 66, 92; roach 15, 19, 26, 32, 46, 52, 56, 57, 65, 66, 92; salmon/trout 57, 92; tench/perch 15, 44, 47, 52, 53, 55, 65, 66, 92; tiddlers 46, 92
Song Dynasty 74
species/hybrids 7
spines 5–6
spread of crucians 14–15; Australia 16; ducks and 16–17; India 16; Ireland 15–16
spring viremia of carp (SVC) 25
St Croix island 72
Sterba, Günther 11
Stewart, Charles 13
summer fish 44–45
Sundadanioninae 25
superpowers 8–10
SVC *see* spring viremia of carp (SVC)
sweetcorn 45, 50
swimfeeder 50, 52

T

taxonomy 25–37
Tench (*Tinca tinca*) 15–16, 26, 44, 52–53, 55, 65–66, 92
tiddlers 46, 92
Tincinae/Tincidae 25, 26
trout 49, 57, 58, 77, 92
Turnbull, Chris 4, 37, 42, 76, 85

U

United States Virgin Islands (USVI) 72
University of Upsal 2, 12
USVI *see* United States Virgin Islands (USVI)

V

verones 15

W

Walker, Richard Stuart 42
Walton, Izaak 74, 84
Wells, A. Lawrence 14, 91
wels catfish (*Silurus glanis*) 84, 91

Wheeler, Alwyne 17, 30–31, 62, 90, 91
wildlife 10, 78, 80, 81, *84*, 85, 88, *89*
Wildlife and Countryside Act 1981 82
Williamson, Henry 76
Willow Pitch VI: Crucian Renaissance (Turnbull)
        4, 37, 42, 76, 87
worms 18, 45–47, *47*

X

xenocypridinae 25

Y

Yarrell, William 12, 31
younger crucians 5

Z

zander (*Stizostedion lucioperca*) *57*, 91
zooplankton 16
Золотой карась (golden crucian), defined 72–73